ANATOMY AND PHYSIOLOGY
FOR NURSES

NURSES' AIDS SERIES

Nurses' Aids Series

ANATOMY AND PHYSIOLOGY FOR NURSES

9th Edition

Sheila M. Jackson SRN, SCM, BTA, RNT
Course Coordinator, Diploma of Applied Science (Nursing),
Riverina College of Advanced Education,
Wagga Wagga, New South Wales, Australia;
formerly Inspector of Training Schools,
General Nursing Council for England and Wales

BAILLIÈRE TINDALL
London Philadelphia Toronto Sydney Tokyo

Baillière Tindall   24–28 Oval Road,
W. B. Saunders   London NW1 7DX

The Curtis Center,
Independence Square West,
Philadelphia, PA 19106–3399 USA

55, Horner Avenue
Toronto, Ontario M8Z 4X6, Canada

Harcourt Brace Jovanovich (Australia) Pty Ltd,
32–52 Smidmore Street, Marrickville, NSW 2204,
Australia

Harcourt Brace Jovanovich (Japan) Inc.
Ichibancho Central Building, 22–1 Ichibancho,
Chiyoda-ku, Tokyo 102, Japan

First published 1939
Ninth edition 1979
   Reprinted 1981, 1982, 1984, 1985, 1986, 1989 and 1990

ELBS edition 1979
(ISBN 0 7020 0738 2)
Sinhala edition (Official Language Affairs, Colombo) 1958
Turkish edition (Turkish Government) 1960
Spanish edition (C.E.C.S.A, Mexico) 1973
Hindi edition (N. R. Brothers, Indore) 1978

Printed in England by Clays Ltd, St Ives plc

Jackson, Sheila M.
   Anatomy and physiology for nurses.—9th ed. (Nurses' aids series).
   1. Human physiology
   I. Title   II. Armstrong, Katherine Fairlie
   612′.002′4613   QP34.5
ISBN 0 7020 0737 4

Contents

Preface

Just under half a century ago, Katherine Armstrong, in the Preface to the First Edition stated her intent 'to provide nurses with a handbook which covers the syllabus of the General Nursing Council for England and Wales in human anatomy and physiology and which shall yet be a small volume at a reasonable price. Stress is laid throughout on the ... approach which encourages the nurse to understand and not merely to memorize the knowledge necessary to pass her examinations, and makes her able to carry out her practical nursing work in an intelligent manner.' In this Ninth Edition, and its predecessor, I have sought to adhere to these basic principles. To meet the needs of today's reader and the new approach to the teaching of anatomy and physiology resulting from new knowledge and from consequent changes in the nurses' syllabus, substantial revision and rewriting were essential.

The earlier chapters introduce briefly some scientific principles which will help in the understanding of the chemistry of the body in later chapters. Here, and throughout the book, S.I. measurements are used, and the terminology adheres to the recommended English equivalent of the Paris *Nomina Anatomica*. In the chapter The Blood, I have incorporated new material on the reticulo-endothelial system, and the section on immunity has been expanded to include details of immune reactions (allergies) and autoimmunity. Further updating of the text involves pulmonary, portal and fetal circulation; the biliary system, its parts and function, in particular, the expansion of the section dealing with the pancreas. Nutrition and metabolism are given special emphasis in one chapter to underline their relationship. The chapter on the nervous system has been expanded and reorganized to facilitate understanding of this complex subject and additional information is included on the sensory system, special senses and sensation from skin and muscle, the meninges and cerebrospinal fluid. A list of Further Readings has been added for those students wishing to gain a little deeper knowledge of certain subjects.

I am extremely grateful for the willing help given by my typist Mrs. K. M. Webb, whose punctuality and encouragement have coerced me into completing the work within the allotted time. My thanks also go to my family for putting up with me during the writing of the book and to the publishers who always give their help so unstintingly.

December 1978   SHEILA M. JACKSON

1 Elementary Physics and Chemistry

The nurse cares for human beings in illness and during vulnerable periods of their lives. In order to do this properly she needs to understand the structure and function of the body and to be able to relate this knowledge to her care of the patient. Each organ in the body plays its part in maintaining the health of the whole, and if one organ is at fault the whole body will be affected. The structure of each part suggests the function, and the function suggests the structure, so that the study of the human body is a logical process of thinking and reasoning, not merely of memorizing.

Terms. *Anatomy* is the study of the structure of the body and *physiology* is the study of its function. Some knowledge of elementary physics and chemistry will help in the study of anatomy and physiology and will also be needed to enable the nurse to care for the patient and to carry out nursing procedures. *Chemistry* deals with the composition of matter and the reactions between various types of matter. *Physics* deals with the behaviour and characteristics of matter, for example whether it gives off heat and light, or conducts electricity.

Matter

Matter is anything that can occupy space. When we refer to a 'space' we usually mean that it contains air, which can be displaced by other forms of matter. For example, an 'empty' bottle is not really empty as it contains air. A container from which the air has been extracted is called a vacuum.

Physical States of Matter

Matter has three physical states: (a) solid, (b) liquid and (c) gas.

A *solid* does not easily alter in either shape or size, e.g. stone, brick.

A *liquid* takes the shape of the vessel it is put into, but does not alter in size, e.g. 500 millilitres (ml) of water has no shape, but it takes the shape of the container it is put into; if it is put into a jug which will hold 1 litre the jug will not be filled.

A *gas* takes the shape and size of the vessel containing it. If the air is extracted from a container and a small quantity of gas is introduced the gas will expand to fill the container. If more gas is introduced the gas in the container is compressed, and the small particles which make up the gas are packed more closely together. More gas could be introduced until the flask bursts. A common example of this can be seen when a bicycle tyre is pumped up. After a few pumps the tyre seems full, but when someone gets on the bicycle the tyre goes flat because there is so little air in it. You can go on pumping until there is so much air in the tyre that when you get on the bicycle your weight does not flatten the tyre. A measured amount of gas in a cylinder will exert a known pressure. As the gas is used up the pressure in the cylinder will drop and it is in this way that the quantity of gas remaining in the cylinder can be gauged. Common examples of gases are oxygen, hydrogen and nitrogen.

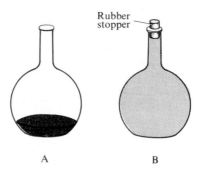

FIG. 1. Glass flasks each containing 100 millilitres of (A) fluid and (B) gas. Note how the gas expands to fill the flask.

Change of State

At normal temperatures iron is solid and water is liquid, but the state of matter can alter. The changes are due to heating or loss of heat. For example, liquid water becomes solid ice when heat is lost, and solid butter becomes liquid oil when it is warmed. The changes can be described as follows.

Melting is the change of a solid into a liquid as a result of heating, e.g. the turning of ice into water.

Evaporation is the change of a liquid into a gas as a result of heating, e.g. the turning of water into steam or water vapour. Gases formed by evaporation are called vapours.

Condensation is the turning of a gas into a liquid as a result of cooling, e.g. the turning of water vapour into water.

Consolidation is the turning of a liquid into a solid as a result of cooling, e.g. the turning of water into ice.

The first two of these changes, melting and evaporation, are due to heating, but the heat which causes this change of state does not cause any rise in the temperature of the matter. Heat is a form of energy and the energy in this case is used up in producing the change of state, so it does not heat the matter. For example, the temperature of melting ice is 0°C (32°F). When the ice melts the water is also at a temperature of 0°C. Water boils at 100°C (212°F). If you leave the kettle on the heat the water will not get any hotter because the added heat is used up in turning the water into water vapour. In the same way heat from the body evaporates the sweat from the skin, excess body heat is used up and the body is cooled. Heat used up in this way is called *latent heat*. In condensation or consolidation latent heat is set free.

Analysis and Synthesis

Earlier it was explained that chemistry is the study of the composition of matter. A chemist tries to split matter up into its various components to see what it is made of. This process is called *analysis*. He may also try to build up matter from its various components. This process is called *synthesis*. Synthetic substances are those which are made by man, e.g. plastics and nylon.

Elements, Compounds and Mixtures. Some substances cannot be split up and these substances are called *elements*. There are over ninety elements known to exist, including oxygen, carbon, nitrogen, iron, silver and gold. Substances which can be split into different elements are of two types: (a) compounds, and (b) mixtures.

A *compound* is a substance made of two or more elements which combine chemically to form a new substance with new properties. For example, hydrogen is a very light gas, used to fill balloons. It is highly inflammable and may explode. Oxygen is a gas which supports combustion but will not itself burn. These two gases in combination make a compound, water, which is a fluid, but it is not inflammable, will not explode, will not support combustion and will not burn. In fact it is very useful for putting out a fire. The elements which form a compound are always present in fixed proportions. For example, water consists of two parts of hydrogen combined with one part oxygen. If the proportions were altered a compound might be formed, but it would not be water. Two parts of hydrogen combined with two parts of oxygen give a compound called hydrogen peroxide which looks like water but has very different properties. Another characteristic of a compound is that it is not easy to separate the elements which compose it. To accomplish this separation a chemical change must take place.

A *mixture* is a substance made of two or more elements mixed together but not chemically combined. Air is a mixture of gases, mainly nitrogen and oxygen, with traces of carbon dioxide and other gases. A mixture has no new properties but simply those present in the elements which form the mixture. Air supports combustion because it contains free oxygen, but it does not support combustion as well as pure oxygen because of the high percentage of nitrogen, which will not support combustion. The elements in a mixture can be separated fairly easily because they are not chemically combined. Oxygen in air will support combustion because it is 'free' oxygen. Oxygen in water is not free and therefore will not support combustion. In a mixture there is no fixed proportion of elements composing it. Three different samples of air may contain three different proportions of oxygen, yet each is a specimen of air.

Common Elements. Among over ninety elements some are very rare, but there are a few common ones which should be known, with their symbols:

Barium	(Ba)	a malleable solid (metal)
Calcium	(Ca)	a solid (metal)
Carbon	(C)	a solid (non-metallic)
Hydrogen	(H)	a gas
Iodine	(I)	a solid (non-metallic)
Iron	(Fe)	a solid (metal)
Magnesium	(Mg)	a solid (metal)
Mercury	(Hg)	a liquid (metal)
Nitrogen	(N)	a gas

Oxygen	(O)	a gas
Potassium	(K)	a solid (metal)
Sodium	(Na)	a solid (metal)

Many of these elements are of interest to nurses. *Barium* is a metal. Its salts are opaque to X-rays and can be used for outlining the digestive tract for diagnostic purposes. *Calcium* is the mineral which makes teeth and bones hard. *Carbon* is rare as an element but is present in many compounds and in all living matter, including food. *Hydrogen* is a light, highly inflammable gas and when mixed with oxygen it forms an explosive mixture. *Iodine* is obtained from green food such as vegetables and is present in large quantities in sea-weed. It is needed by the body to make the secretion of the thyroid gland. *Iron* is a metal and a small quantity is present in the body. It is necessary for the manufacture of the red cells in the blood and is important for health. Lack of iron is the most common cause of anaemia. *Magnesium* is found in small quantities with calcium in the bones and teeth. *Mercury* is poisonous to the body, but is useful in thermometers because it expands as it becomes warmer. *Nitrogen* is a gas which neither burns nor supports combustion; it is present in all living matter and must be present in our food supply. *Oxygen* is a gas which will not itself burn but which will support combustion; combustion is also called oxidation. *Potassium* is present in all living matter, particularly in plants. In the human body it is present in tissue cells. *Sodium* is also present in all living matter, but particularly in the animal world. In the body sodium is chiefly in the form of sodium chloride. There are approximately $4 \cdot 5$ grams (g) (1 teaspoon) of sodium chloride in $0 \cdot 5$ litre of body fluid (9 g to 1 litre = $0 \cdot 9$ per cent solution).

All compounds and mixtures are made of elements such as these. The body is made of very complicated chemical compounds built up from a small number of elements, so it is necessary for the student of anatomy and physiology to know a little about elements and compounds.

The Structure of Matter

All matter is made up of tiny particles called *molecules* which are so small that they cannot be seen with an ordinary microscope. Molecules are held together only by the attraction of one molecule to another, in the same way that a needle is held to a magnet. It is perhaps rather difficult to accept the fact that an iron bar is made of molecules which do not even adhere to each other but are only held together by mutual attraction. It is, however, an important concept which must be remembered. The force of attraction is called *cohesion* and must be distinguished from adhesion. A molecule is the smallest particle which can exist alone but which still has all the properties of the substance. For example, a molecule of water has all the properties of water which were mentioned earlier.

A molecule can be divided into smaller particles called *atoms*, but an atom cannot exist alone. A molecule of an element consists of atoms of that element, e.g. a molecule of oxygen consists of two atoms of oxygen (written O_2). A molecule of a compound consists of atoms of the different elements present in the compound, e.g. a molecule of water consists of two atoms of hydrogen and one atom of oxygen (written H_2O). A molecule of carbon dioxide consists of one atom of carbon and two atoms of oxygen (CO_2) and a molecule of glucose, which is a simple form of sugar, consists of six atoms of carbon, twelve atoms of hydrogen and six atoms of oxygen ($C_6H_{12}O_6$). These are examples of very simple chemical compounds.

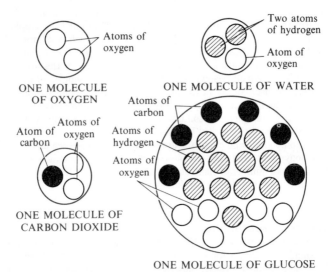

FIG. 2. Diagram to represent molecules of oxygen, water, carbon dioxide and glucose.

In the more complicated compounds which form the human body many elements may be present, and the number of atoms of any one element may run into hundreds. Although all molecules are very small they vary considerably in size and complexity.

Atomic Number and Atomic Weight. The word atom means indivisible and the name was given when scientists thought that the atom could not be divided. Now, however, more is known about the structure of an atom and most people have heard about 'splitting the atom' and its connection with atomic energy. An atom consists of even smaller particles of three different types, *protons* which carry a

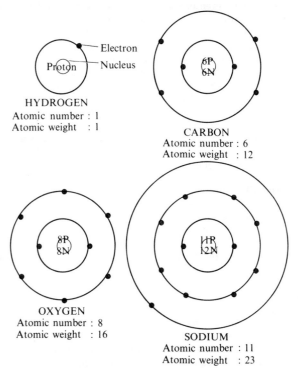

HYDROGEN
Atomic number : 1
Atomic weight : 1

CARBON
Atomic number : 6
Atomic weight : 12

OXYGEN
Atomic number : 8
Atomic weight : 16

SODIUM
Atomic number : 11
Atomic weight : 23

FIG. 3. Diagram of the structure of atoms of hydrogen, carbon, oxygen and sodium. P = proton; N = neutron.

positive electrical charge, *electrons* which carry a negative electrical charge and *neutrons* which are neutral. The protons and neutrons are massed together to form the nucleus of the atom, while the electrons whirl around the nucleus in one or more orbits known as shells. Different elements have different numbers of protons and electrons forming their atoms and the number of protons is equal to the number of electrons. This number is called the *atomic number*. Hydrogen has one proton in its nucleus and one electron circling round it and its atomic number is one. It has no neutron. Carbon has six protons and six neutrons in its nucleus and six electrons, arranged in two shells, circling round it, so its atomic number will be six. Sodium has eleven protons and twelve neutrons in the nucleus, and eleven electrons, in three shells, circling round it, so the atomic number is eleven. These numbers do not have to be memorized because they can be obtained from tables. The presence of neutrons in the nucleus does not affect the atomic number.

The weight of a neutron is roughly equal to the weight of a proton and the number of neutrons added to the number of protons gives the *atomic weight*. The weight of an electron is so small that it is not significant. Hydrogen has an atomic weight of one as it has one proton and one electron but no neutron. Carbon has an atomic weight of twelve as it has six protons and six neutrons in its nucleus. Sodium has an atomic weight of twenty-three as it has eleven protons and twelve neutrons in its nucleus.

Ions and Electrolytes. Atoms do not remain static as electrons may be split off from one atom and gained by another. Atoms are constantly gaining and losing electrons. When a substance is in solution in a fluid and an electron splits from an atom of the substance, an *ion* is formed. Substances which dissociate in this way are known as *electrolytes* because they carry small electric charges. Electrolytes are present in body fluids and are essential to life.

Gas Pressure

Earlier it was stated that one molecule is held to another by mutual attraction. The molecules of a solid are strongly attracted to one another so a solid does not easily alter in size or shape. The molecules of a liquid are less strongly attracted to each other so their positions can be altered more easily and the liquid takes the shape of the container. The molecules of a gas are attracted to each other very little so that when gas is put into a vacuum container the molecules separate from one another and fill the flask. If more gas is introduced the molecules become more densely packed and the gas is said to be under pressure. The pressure of gases is important in their use in the body. Unless oxygen is under sufficient pressure in the lungs it will not be transferred to the blood for transport round the body and the body tissues will suffer from lack of oxygen. On a high mountain the *pressure* of the air is so low that man cannot live without an extra supply of oxygen, though a candle will burn because there is sufficient quantity of air even though the pressure is reduced. In an airless room where there is reduced quantity of oxygen the candle will not burn, though man can continue to live as the oxygen present is still under pressure.

Solutions and Emulsions

When a lump of sugar is put into a cup of hot water the sugar dissolves and a *solution* of sugar is present in the cup. The molecules of sugar mix with the molecules of water until the sugar is evenly distributed throughout the fluid. The process can be speeded up by stirring, but the even distribution of sugar will occur eventually without any stirring. A measured quantity of fluid will only dissolve a measured quantity of another substance and when the fluid has dissolved as much as possible it is called a *saturated solution*. Fluids can dissolve solids, liquids and gases. Oxygen is dissolved in water, which enables fish to live in it. Heating a fluid drives off the dissolved gas, which appears at the surface as bubbles as the fluid gets hot. Water is the most common solvent and can dissolve a wide variety of substances, but substances which cannot be dissolved in water may be dissolved in other liquids. For example, water will not dissolve oil, but oil will dissolve in spirit. The fact that water is a common solvent makes it important in the body, two-thirds of which is water. Most of this water is in the cells which make up the body, some of it surrounds the cells, which can only live in a salt solution, called saline, and some is in the circulating blood and lymph.

An *emulsion* of oil is different from a solution. If oil is mixed with water the oil will rise to the top when stirring is stopped and will form a layer of pure oil on top of the water. If the oil is mixed with a solution of sodium carbonate (washing soda) an emulsion will be formed. The oil will be broken up into little droplets which will be suspended in the fluid and will not run together again. When allowed to stand the oil will still rise to the top, but the fat droplets will remain separated. Milk is a natural emulsion and the droplets of fat which form the cream can easily be seen under a microscope.

Acids and Alkalis

Substances may be either acid or alkaline in reaction, or they may be neutral (neither acid nor alkaline). The simplest agent used to detect the reaction of a substance is *litmus* which can be obtained as a fluid, but is conveniently used in the form of litmus paper. Blue litmus paper will turn pink when in contact with acid. Pink litmus paper will turn blue when in contact with an alkali. Neutral substances turn both pink and blue litmus paper into a purple colour. Every living cell must have the correct reaction if it is to survive. The blood and body tissues are slightly alkaline in reaction and vary little throughout life. If too much acid or too much alkali is present in the body illness will occur, and if there is an alteration in the reaction of the blood or body tissues death will follow. It is necessary to know the degree of acidity or alkalinity of certain substances and for this purpose a universal indicator is used. It changes colour from bright red, indicating a strong acid, to purple, which indicates strong alkali. These colour changes are spoken of as the pH of the substance being tested, and the scale runs from pH 1 for the strong acid to pH 14 for the strong alkali. Neutral is pH 7 and the pH of the blood is 7·4.

Pure water is neutral in reaction and as previously stated consists of molecules composed of two atoms of hydrogen and one atom of oxygen. Some of the molecules are broken up into smaller particles, e.g. hydrogen ions (H ions) or hydroxyl ions (OH ions). H ions have a positive electrical charge because the atoms have lost an electron and OH ions a negative electrical charge. In pure water the H ions and the OH ions are equal in number and the reaction is neutral. In some solutions H ions may outnumber OH ions and the reaction of the solution will be acid. If the OH ions outnumber the H ions the reaction will be alkaline. The pH shows the concentration of hydrogen ions (the '*H ion concentration*'), strong acids, e.g. hydrochloric acid, having a high H ion concentration, and strong alkalis, e.g. caustic soda, the highest concentration of hydroxyl ions (OH ions). Some substances, which slow down changes in reaction, are called '*buffer substances*'. In an alkaline solution such as body fluids, buffer substances neutralize acids or any excess of alkalis which may be added to it or produced in it. Examples of buffer substances are sodium, potassium and protein which serve as buffers against acid substances in the body and carbonic acid, lactic acid and fatty acids which serve as buffers against any excess of alkaline bases.

Mineral Salts

A mineral salt is a substance made by the action of acid on a mineral. The mineral is called the *base* of the salt, and the scientific name shows the base and the acid which together form the salt, e.g. sodium chloride is made by the action of hydrochloric acid on sodium. A salt may be acid, alkaline or neutral in reaction. A strong acid acting on a strong alkali forms a neutral salt; a strong acid acting on a weak alkali forms an acid salt; a weak acid acting on a strong alkali forms an alkaline salt. The various salts present in the body differ only slightly in reaction.

Measuring the Amount of a Substance

Earlier it was stated that the number of neutrons added to the number of protons in an atom gives its *atomic weight*. Hydrogen has an atomic weight of 1 (as it has one proton and no neutrons) and oxygen an atomic weight of 16 (as it has eight protons and eight neutrons). A molecule of water, which consists of two hydrogen atoms and one oxygen atom (H_2O), will therefore have a weight of 18. This is known as its *molecular weight*. Understanding atomic and molecular weights is important in understanding how the amount of a substance is measured.

The amount of a substance can be expressed in terms of its mass, but in physiology and biochemistry it is more important to know *how many particles* (atomic, ionic or molecular) of the substance are present rather than their total mass. However, since even tiny amounts of a substance will contain many millions of particles, it would be very cumbersome to use such huge numbers. Therefore a unit defined as a standard large number of particles was devised. The number of particles chosen was the number of particles in that amount of a substance that has a mass equal to its molecular or atomic weight in grams; this number will always be 6.023×10^{23} and is referred to as Avogadro's number. For example, the atomic weight of carbon is 12, therefore 12 g of carbon will contain 6.023×10^{23} atoms of carbon. The molecular weight of water is 18, therefore 18 g of water will also contain 6.023×10^{23} molecules of water. The amount of substance containing this number of particles is called a *mole*, the symbol for which is mol.

The mole can be used to indicate the concentration of a substance in a solution. If one mole of a substance is added to one litre of water, the concentration of the substance in the resulting solution will be 1 mol/l. Because the mole is a measure of the number of particles, one mole of a molecule which breaks up into ions when added to water will produce one mole of each of the ions produced. For example, hydrochloric acid (HCl) breaks up into hydrogen ions (H^+) and chloride ions (Cl^-) when added to water. One molecule of HCl will produce one H^+ and one Cl^-. Therefore 6.023×10^{23} molecules of HCl (one mole of HCl) will produce 6.023×10^{23} H^+ (one mole of H^+) and 6.023×10^{23} Cl^- (one mole of Cl^-).

The mole is often too large a unit to describe the concentrations of substances encountered in clinical medicine. Therefore a prefix is used, as for other units. The prefix commonly used is milli-, one millimole (mmol) being 1000 times smaller than a mole. For example, the concentration of sodium chloride (NaCl) in the blood is approximately 0.15 mol/l, but this is usually written as 150 mmol/l. A solution made up to have this same concentration of NaCl (i.e. 150 mmol/l) is known as *isotonic* or *physiological saline*.

Moles cannot be used for measuring the concentration of protein substances because the molecules are so complex that their molecular weight cannot be determined accurately. Such substances are therefore expressed in terms of their mass, i.e. in grams per litre or in milligrams (mg) per 100 ml. It is rarely necessary to convert from one to the other but if it should be the formula is:

$$\text{concentration in mmol/l} = \frac{\text{concentration in mg/100 ml} \times 10}{\text{molecular weight of substance}}$$

The concentrations of dissolved gases, e.g. in the blood, are often stated in terms of pressure rather than moles or grams because pressures are easier to measure.

Diffusion

If two gases of different composition come into contact, intermingling of the gases takes place until the composition of both is the same. For example, the air we breathe out contains more carbon dioxide and less oxygen than the air we breathe in, but the expired air does not float around the room as a portion of different air. The air in the whole room will eventually contain less oxygen and more carbon dioxide overall. This process is known as *diffusion*. The term diffusion is also used to describe the passage of small molecules of acids and salts through a semipermeable membrane. A membrane may be described as permeable when it allows substances in solution to pass through; as impermeable when no fluid can pass through, and as semipermeable when water and small molecules in solution can pass through but not larger molecules. If a strong

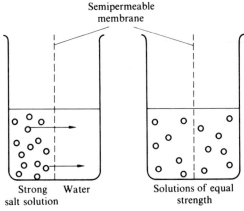

Fig. 4. Diagram to show the diffusion of liquids.

solution of salt is separated from a weaker solution by a semipermeable membrane the salt molecules will pass through the membrane from the strong solution to the weaker one until both solutions are of equal strength.

Osmosis

If a substance with large molecules, e.g. sugar, is made into a solution and is separated from a weaker sugar solution by a semipermeable membrane only water will pass through the membrane, from the weak solution to the strong solution, because the molecules are too large to pass. This passage of water across a semipermeable membrane is called *osmosis*.

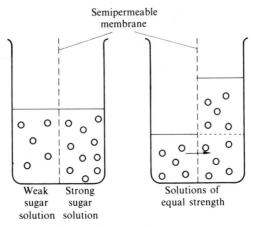

FIG. 5. Diagram to show osmosis.

Specific Gravity

Specific gravity is the weight of a known volume of liquid divided by the weight of an equal volume of pure water.

$$\text{Specific gravity} = \frac{\text{Weight of substance}}{\text{Weight of equal volume of water}}$$

The specific gravity of pure water is expressed as 1·000. The specific gravity of urine ranges between 1·010 and 1·020, and that of blood is about 1·055.

Organic Matter and Inorganic Matter

Organic matter is that which is living or has been alive, e.g. wood or coal. Inorganic matter is that which is not living and never has been alive, e.g. water or iron. Inorganic substances can be built up into organic compounds by living organisms.

2 Characteristics of Living Matter

In the last few years scientists have come very close to an understanding of the mechanisms of life, and this new understanding has shown an underlying unity in all living things. For example, the fundamental molecules and mechanisms in a cabbage are the same as those in a man, and it is becoming possible to give some answers to questions such as 'What is life?' All life is made up from several dozen chemicals put together in similar ways and doing approximately the same job, no matter what organism they come from. And life has another unity. All living matter is made up of small units called cells. Some organisms, such as bacteria, consist of one single cell; others, like man, consist of many hundreds of millions of cells, all functioning together to make a complete whole.

The Characteristics of Living Matter

All living cells, however simple, have certain characteristics which are always present. These characteristics are:

1. Activity
2. Respiration
3. Digestion and absorption of food
4. Excretion
5. Growth and repair
6. Reproduction
7. Irritability

Activity

This is the most striking characteristic of living matter. It is more apparent in the animal world than in the vegetable world, as the animal must move about in search of food, but even in plants the buds can be seen to break out in the spring, and with a microscope activity becomes as obvious in the plant as in the animal. There can never be any activity without energy. Cars are driven by energy derived from petrol, trains are generally moved by electrical energy, and in living matter the energy is also obtained by burning fuel. The fuel for the human body is the food eaten, particularly carbohydrates and fats. Oxygen is also necessary for the combustion of fuel, and living things obtain their oxygen from the air, or water, in which they live. Combustion of fuel also produces waste products, such as carbon dioxide and water, which must be disposed of.

The combustion of fuel in living matter produces some energy for work and some energy in the form of heat. The body is economical in that it produces more work energy and less heat for each unit of food consumed, but the heat produced by combustion is not altogether wasted, for some heat is needed. The body can remain healthy only within a narrow temperature range of 36° to 37·5°C (97° to 99·5°F) and any heat produced in excess of this must be got rid of if life is to continue.

Respiration

All living matter requires oxygen and gives off carbon dioxide. The oxygen is required for combustion, or oxidation, of food, and carbon dioxide is the waste product of combustion. This process of taking in oxygen and giving off carbon dioxide is called respiration and it is continuous throughout life. The amount of oxygen required and the quantity of carbon dioxide given off vary with the amount of activity taking place. During sleep the human body requires comparatively little oxygen, but it would require much more during strenuous exercise such as playing football or climbing a mountain.

Digestion and Absorption of Food

All living matter requires food. Some food can be absorbed as it is, but most of it must be broken down into substances which have smaller and more simple molecules, before it can be absorbed. The process of breaking down complex foods into simpler substances is termed digestion. It is brought about by enzymes, which are themselves protein substances and which act upon the food, preparing it for absorption.

Excretion

All living matter produces waste products which are of no further use and must be got rid of. If allowed to accumulate, waste products will interfere with the processes of life. In the human body these waste products include carbon dioxide from the lungs, sweat from the skin and urine from the kidneys.

Growth and Repair

Living matter is able to build up new living matter, similar to itself, from food which is taken in. Protein foods include meat, cheese and milk, and these provide the building material for the body. Because the body is continually active its parts are constantly being worn out and they are made good by the building up of new living matter. In childhood, growth and repair are taking place simultaneously. In old age the building up does not keep pace with the breaking down, so weakness begins to appear and will eventually cause death unless illness supervenes. It is this ability to grow and to repair its own tissues, and also the power of reproduction, that makes living matter different from man-made substances.

Reproduction

All living matter can reproduce its own kind. In the simplest forms of life reproduction is a very simple process, involving the splitting of the parent cell into two. In animals, female cells called ova are produced in the ovaries, and sperms, or male cells, are produced in the testes. The sperm cell must fertilize the ovum before reproduction can take place.

Irritability

All living matter is able to respond to a stimulus, and in higher animals this means the power to be aware of the environment and to respond to it. Irritability is more marked in animals, but it can be observed in plants. If a bulb is planted upside down the roots will still grow deep into the soil and the leaves into the air, or if a plant is placed in a dark corner it will grow tall and thin as it seeks the light. This indicates that even plants are aware of the environment. In animals, irritability allows the awareness of danger, the recognition of that which is useful, e.g. food and water, and the exercise of free will.

All this is what is meant when the physiologist talks about life. Living matter is that which has all the characteristics mentioned above, and all these characteristics are the result of chemical changes. All living matter has the ability to produce protein substances called *enzymes* which in turn control and produce chemical changes within the cell. Enzymes cause other substances to combine with one another, or to split up, though they themselves do not enter into the reaction. For example, complex starches combine with water and split up into simple sugars in digestion; in combustion sugar splits up into carbon dioxide and water. All these changes are brought about by enzymes.

3 The Structure of Living Matter

The cell is the unit of living matter, the fundamental building block of life, and all cells are very much alike, regardless of their origin.

Structure of a Cell

All cells are made of a substance called *protoplasm*, which is jelly-like, opaque and colourless and consists of water and other substances in solution. These other substances include: (a) organic and inorganic salts, (b) glucose and (c) nitrogenous substances.

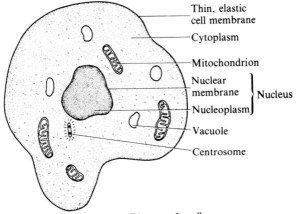

Thin, elastic cell membrane

Cytoplasm

Mitochondrion

Nuclear membrane } Nucleus

Nucleoplasm }

Vacuole

Centrosome

Fig. 6. Diagram of a cell.

The word *cytoplasm* is commonly used to describe the protoplasm which forms the bulk of the cell, the prefix 'cyto' being derived from the Greek word for cell. The cytoplasm is constantly being worn out, broken down and replaced by fresh protoplasm, built up by the cell from the food it takes in. The cytoplasm contains molecules of protein known as ribonucleic acids (RNA) which act as messengers carrying information out from the nucleus to the cytoplasm.

The *cell membrane* surrounds the cytoplasm and is a semi-permeable membrane. The tiny 'pores' in the membrane allow minute molecules to pass into the cell and out of it. The membrane is also thin and elastic and it yields to pressure, so that the cell is able to change its shape.

The nucleus is a dense body within the cell, contained within the *nuclear membrane*. The protoplasm inside the nuclear membrane is known as the *nucleoplasm*. The characteristic compounds of the nucleus are deoxyribonucleic acids (DNA), which contain the genetically inherited information required for the maintenance of the cell. The nucleoplasm stores the information necessary for the cell to grow and to divide into two daughter cells. This information is stored in the *genes*, which are strung together to form *chromosomes*.

Chromosomes are normally only visible under the microscope when the cell is about to divide. Genes are composed of DNA.

The *mitochondria* are the power stations of the cell. They are responsible for converting the food ingested by the cell into energy.

Vacuoles are clear spaces within the cytoplasm, which contain waste materials, or secretions, formed by the cytoplasm.

The *centrosome* is a small rod-shaped body near the nucleus and is important in cell division. It is surrounded by a radiating thread-like structure and contains two centrioles.

Cell Reproduction

Simple Fission

The nucleus plays an important part in reproduction. In *simple fission* the nucleus in a single cell becomes elongated and then divides to form two nuclei in one cell. The cytoplasm then divides between the nuclei to form two daughter cells each with its own nucleus (Fig. 7). These smaller cells will take in food and grow and when they are full size they will divide to form a total of four cells.

Mitosis

In the more complex forms of life, and this includes the cells in the human body, cell division is a more complicated process, which is called *mitosis* (Fig. 8). There are seven stages:

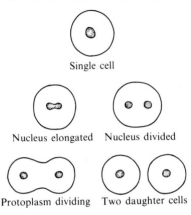

Single cell

Nucleus elongated   Nucleus divided

Protoplasm dividing   Two daughter cells

FIG. 7. Diagram illustrating simple fission.

1. The centrosome divides into two and each moves away from the other, though they are still attached by thread-like structures. This stage is called *prophase*.

2. The nuclear material forms dark thread-like structures which are the chromosomes. There are forty-six in human cells, though the number differs in other species.

3. The nuclear membrane disappears and the chromosomes arrange themselves around the centre of the cell. They appear to be attached to the thread-like structure of the centrosomes, which are by now at either end of the cell. These two changes are known as *metaphase*.

4. The chromosomes divide longitudinally into two equal parts.

5. The two groups of chromosomes move away to either end of the cell and arrange themselves around the centrosomes. The thread-like structures joining the centrosomes now divide. These two changes are called *anaphase*.

6. The cell body gets narrower round the centre. The thread-like structures disappear and two nuclear membranes reappear. This is called *telophase*.

7. The cell divides and the chromosomes disappear into the nucleus. The daughter cells will in turn grow and will reproduce by mitosis.

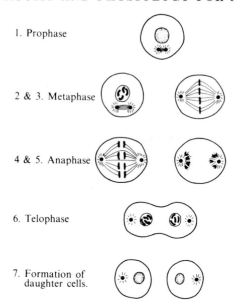

1. Prophase

2 & 3. Metaphase

4 & 5. Anaphase

6. Telophase

7. Formation of
daughter cells.

FIG. 8. Diagram illustrating mitosis.

Meiosis

Reproduction in higher animals, including man, depends on the fusion of a spermatozöon from the male parent and an ovum from the female. These reproductive cells are also called *gametes*. Each gamete must receive half as many chromosomes as normal cells so that when they unite in fertilization the normal number of chromosomes is obtained. As the sex cells mature two processes of cell division occur; the first is mitosis, in which each daughter cell receives a complete set of chromosomes. Following this a two-stage cell division occurs which is peculiar to reproductive tissue and is called *meiosis*. The first division is similar to mitosis and gives rise to two daughter cells each containing the original number of chromosomes; the second division follows the first very quickly and results in four gametes each containing half the number of chromosomes. In man normal body cells have forty-six chromosomes (in twenty-three pairs) while each gamete has only twenty-three single chromosomes. Following fusion of the gametes the resulting cell, called a *zygote*, has the original forty-six chromosomes (in twenty-three pairs). Cell division of a zygote is by mitosis and the resulting multi-celled organism is called an *embryo*.

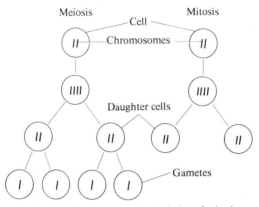

Fig. 9. Diagram comparing meiosis and mitosis.

The chromosomes consist of a chain of genes strung together, and it is the genes which pass on the characteristics of the parent cell, so that daughter cells are always similar to parent cells. In this way the physical and intellectual characteristics of the children are derived from the parents, including such characteristics as hair colour, height, degree of intelligence and many others. In any pair of genes one will exert a stronger influence than the other. The stronger gene is termed *dominant* and the weaker one *recessive*. The characteristics which are hereditary depend upon the dominance of the genes.

Sex Determination. One pair of chromosomes from the father and one pair from the mother are the sex chromosomes which will determine the sex of the child. In the female the sex chromosomes are the same and are called XX. In the male they are different and are called XY. One chromosome from each pair will determine the sex of the child. If the child has an X chromosome from the mother and an X chromosome from the father it will be a girl (XX). If the child has an X chromosome from the mother and a Y chromosome from the father it will be a boy (XY).

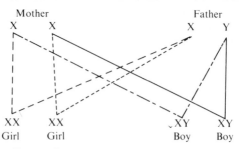

Fig. 10. Diagram to show sex determination.

Unicellular Organisms

These are organisms in which one cell forms a complete individual;
examples are bacteria and amoebae. The cell carries out all the
characteristic activities of living matter. Movement of the amoeba is
performed by the flow of protoplasm within the cell. A projection is
pushed out in the direction in which the movement is to be made
and the protoplasm streams slowly into the projection until the
whole cell occupies a new position. The projection is called a *pseudo-
podium* or false foot and this type of movement is called *amoeboid
movement*. In the human body it is carried out by the white blood
cells.

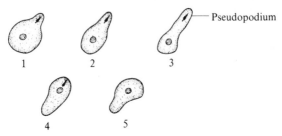

Fig. 11. Diagram to show amoeboid movement.

Unicellular organisms are also able to ingest food. Two pseudo-
podia are put out and the food is surrounded. Enzymes are then
poured out and gradually the food is digested and absorbed, until
only a small quantity of waste material is left in the vacuole. The
cytoplasm then flows back so that the waste material is left behind.

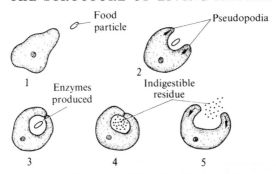

FIG. 12. Ingestion and digestion of food particles.

Multicellular Organisms

Multicellular organisms consist of many cells. Each cell is living and requires food, oxygen, water, suitable temperature and correct pH, but it is dependent on other cells to supply it with the necessities of life in return for the specialized work it carries out. For example, the cells of the lungs take in oxygen, and the cells of the digestive tract absorb food. Each cell develops in a specialized manner so that it can carry out its task satisfactorily. This specialized development is known as the differentiation of cells. Groups of specialized cells, developed for a particular purpose, form the tissues of the body, e.g. muscle tissue for movement and bone for support.

The multicellular organism in which we are particularly interested is the human body, a complex arrangement of millions of cells, but still having the basic characteristics of all living matter.

4 The Tissues

The body consists of countless cells developed to form various different types of tissue. It originates from a single typical cell, the egg cell or ovum, which is composed of protoplasm and contains a nucleus. After fertilization this cell multiplies and forms a ball of cells which, by the process of differentiation, develop into all the various tissues required to form the different organs and parts of the body.

In the very early stages the ball of cells is divided into three layers. The outer layer is called the *ectoderm*, and from it the outer part of the skin develops with its nails, hair follicles and sweat glands and other epithelial tissues including the mucous membrane lining the nose and the mouth and the enamel covering the teeth. The nervous system also originates in the ectoderm. The middle layer is called the *mesoderm* and from it muscle, bone and fat develop and some of the internal organs, including parts of the cardiovascular system. The inner layer is called the *entoderm* and from it the lining of most of the alimentary and respiratory tracts develop.

A *tissue* consists of cells and the products of cells specially developed for the carrying out of a special function. In the body there are four main types of tissue (Table 1):

> Epithelial tissue or epithelium
> Connective tissue
> Muscular tissue
> Nervous tissue

Epithelial Tissue

Epithelial tissue provides covering and lining membranes for the free surfaces inside and outside the body and is the tissue from which the glands of the body are developed. Epithelia protect under-

Table 1
Classification of Tissues

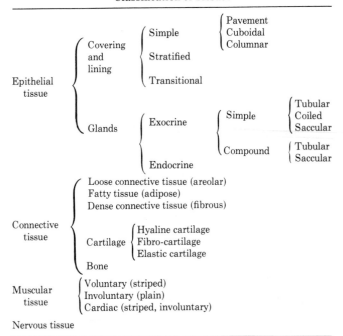

Epithelial tissue	Covering and lining	Simple	Pavement, Cuboidal, Columnar
		Stratified	
		Transitional	
	Glands	Exocrine	Simple: Tubular, Coiled, Saccular; Compound: Tubular, Saccular
		Endocrine	
Connective tissue	Loose connective tissue (areolar); Fatty tissue (adipose); Dense connective tissue (fibrous); Cartilage: Hyaline cartilage, Fibro-cartilage, Elastic cartilage; Bone		
Muscular tissue	Voluntary (striped); Involuntary (plain); Cardiac (striped, involuntary)		
Nervous tissue			

lying tissue from wear and tear, but must continually be renewed as needed. Some cells are specially developed to be able to absorb substances; these linings are only one cell in thickness and often have a specialized surface called a 'brush border'. Some epithelial tissue, particularly glandular tissue, has the ability to secrete substances manufactured within the tissue. Epithelia do not contain blood vessels, the nearest being in the underlying connective tissue, which may be some distance away. The cells are arranged on a 'basement membrane' which plays a part in binding them together.

Covering and Lining Epithelia
Covering epithelia may be classified according to the arrangement and shape of their cells.

Simple epithelium is composed of a single layer of cells attached to a basement membrane; it is very delicate and is found where there is little wear and tear.

1. *Simple pavement epithelium* is composed of flat cells, which form a smooth lining. These may be found lining the blood vessels and forming the peritoneum.

2. *Simple cuboidal epithelium* has cells shaped like cubes and is found covering the ovary.

3. *Simple columnar epithelium* is composed of taller cells, packed onto a basement membrane. It is found where wear and tear is a little greater, for example, lining the stomach and intestines. According to the functions performed columnar epithelium may be modified.

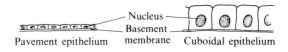

Pavement epithelium — Nucleus — Basement membrane — Cuboidal epithelium

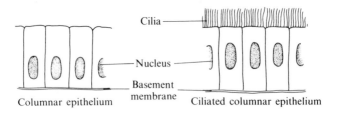

Columnar epithelium — Cilia — Nucleus — Basement membrane — Ciliated columnar epithelium

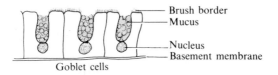

Goblet cells — Brush border — Mucus — Nucleus — Basement membrane

FIG. 13. Diagram showing types of simple epithelium.

Ciliated columnar epithelium has microscopic hair-like processes projecting from the free surface of the cell. The cilia move together with a wave-like motion and mucus and other particles are moved along. This type of tissue is found in the respiratory tract.

Goblet cells are cells which secrete mucus, the mucus collecting in the cells until the cytoplasm becomes distended.

A *brush border* is found on cells specialized for absorption. These have minute finger-like projections which increase the area through which absorption can occur. They are found in the small intestine.

Stratified epithelium is made up of many layers of cells. The deepest cells, called the *germinal layer*, lie on the basement membrane and are columnar. As they divide, and this occurs frequently, the parent cells are pushed nearer the surface and become flattened. The cells on the surface are rubbed off and are continually replaced from below. If the surface of the epithelium is dry, as on the skin, the surface cells die because the blood supply is below the basement membrane, and the scaly surface which develops, called *keratin*, makes a waterproof layer. If the surface is moist, as in the mouth, the cells survive until they are rubbed off, so keratin is not formed.

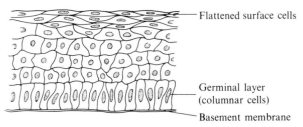

Flattened surface cells

Germinal layer
(columnar cells)

Basement membrane

FIG. 14. Diagram illustrating stratified epithelium.

Transitional epithelium is like stratified epithelium but the surface cells, instead of being flattened, are rounded and can spread out when the organ expands. It is found lining organs which must expand and must be waterproof, e.g. the bladder.

Glands

Glands develop from epithelial tissues and have the ability to manufacture substances from materials brought by the blood. These substances are called the secretions of the gland. For example, blood contains sodium chloride and the gastric glands can make from this the hydrochloric acid found in the gastric juice, although to obtain hydrochloric acid from sodium chloride in the laboratory is a difficult process. Glands are well supplied with blood vessels which supply the cells with the materials necessary for making secretions. Glands are of two types, exocrine and endocrine.

Exocrine glands pour their external secretions out through a duct. The secretions of many of these glands contain *enzymes*, which are chemical substances produced by the cells of the gland. These enzymes produce chemical changes when they come into contact with specific substances, but do not themselves enter into the reaction.

1. *Simple glands* have one duct leading from a single secretory unit.

Simple tubular glands are found in the walls of the small intestine, and the stomach.

Simple coiled glands pour sweat onto the surface of the skin.

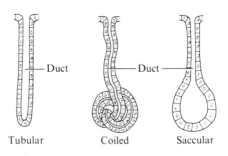

Tubular   Coiled   Saccular

SIMPLE GLANDS

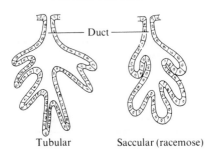

Tubular   Saccular (racemose)

COMPOUND GLANDS

Fig. 15. Types of exocrine glands.

Simple saccular glands, known as sebaceous glands, secrete a substance called sebum which lubricates hair and skin.

2. *Compound glands* have several secretory units pouring their secretions into a number of small ducts which unite to form a larger duct.

Compound tubular glands are found in the duodenum.

Compound saccular glands are also called *racemose glands*, and examples are the salivary glands in the mouth.

Endocrine glands pour their internal secretions directly into the blood stream. These secretions are called *hormones*. Examples of endocrine glands are the pituitary gland in the skull and the thyroid gland in the neck.

Connective Tissue

Connective tissue is tissue which supports and binds together all other tissues. There are many varieties of connective tissue, which differ greatly in appearance, though there is similarity in their connective function and in the fact that all originate from primitive cells called *mesenchyme cells*, which themselves originate in the mesoderm. Connective tissue consists of cells, intercellular substance called matrix, and fibres. Matrix and fibres are non-living products of the cells, which form the supporting material of the body. The fibres are of two main types, collagenous and elastic.

FIG. 16. Collagenous fibres.

Collagenous fibres originate from cells called *fibroblasts*, which secrete a substance which becomes *collagen*. These coarse fibres occur in wavy bundles and will stretch only a little without tearing.

Elastic fibres are fine branching fibres which are highly elastic.

In the body a connection sometimes needs to be firm and unyielding and sometimes a degree of elasticity is required. For example, the fibrous layers surrounding organs must be somewhat elastic to allow for swelling when the part is engorged with blood,

Fig. 17. Elastic fibres.

but the fibrous tendons which join muscle to bone must be inelastic, since if they were elastic the tendon would stretch when the muscle contracted and the bone would not move. There are five main varieties of connective tissue.

Loose Connective Tissue

This type of tissue is also called *areolar tissue* and it consists of a loose network of both collagenous and elastic fibres with small scattered groups of fat cells and some fibroblasts. Some blood vessels and nerves will be found in the tissue but these are not very numerous. Areolar tissue forms a transparent skin, as thin as tissue paper, but very tough, and it is found between and around the organs of the body.

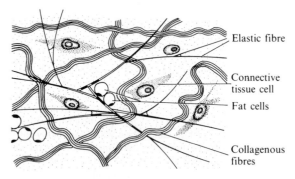

Fig. 18. Loose connective tissue.

Fatty Tissue

This is also known as *adipose tissue* and it is similar to areolar tissue but the spaces of the network are filled in with fat cells. Fat cells contain a globule of fat, which pushes the cytoplasm and the nucleus to the edge of the cell. Adipose tissue is useful because it forms a food reserve on which the body can draw in time of need; it helps retain body heat because it is a poor conductor of heat and it protects delicate organs such as the eye and the kidney.

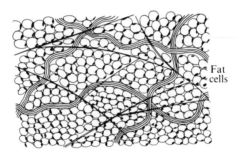

Fat cells

FIG. 19. Fatty or adipose tissue.

Dense Connective Tissue (Fig. 20)

This tissue is also called *fibrous tissue* and it consists chiefly of bundles of collagenous fibres between which the fibroblasts lie. It is very strong compared to loose connective tissue. The fibres may be arranged regularly, in parallel bundles, as in tendons or ligaments, or irregularly, with the fibres running in different directions, as in the sheath enclosing muscles, which is called *fascia*.

Cartilage

Cartilage consists of cells, called *chondrocytes*, separated by fibres. It has no blood vessels, so the cells obtain their nourishment by diffusion of fluid through the intercellular substance. Cartilage is very tough but it is also pliant. There are three types of cartilage:

1. *Hyaline cartilage* consists of chondrocytes embedded in an apparently structureless matrix, which is glassy in appearance, and which has very fine collagen fibres running through it. It is found in the trachea and covering the ends of bones at a joint.

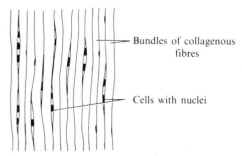

Bundles of collagenous
fibres

Cells with nuclei

Fig. 20. Diagram of dense connective (fibrous) tissue.

2. *Fibro-cartilage* contains more collagen fibres than hyaline cartilage and is therefore stronger. It is found between bones forming slightly movable joints, e.g. between the bodies of the vertebrae.

3. *Elastic cartilage* contains numerous elastic fibres embedded in the matrix and can be found in the auricle of the ear and the epiglottis.

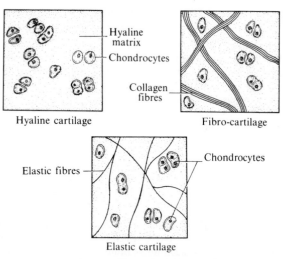

Hyaline
matrix

Chondrocytes

Collagen
fibres

Hyaline cartilage

Fibro-cartilage

Chondrocytes

Elastic fibres

Elastic cartilage

Fig. 21. Types of cartilage.

Bone. Bone is a specialized type of cartilage in which the collagen has been impregnated with mineral salts, chiefly calcium. The collagen fibres make the bone tough and the mineral salts make it rigid, so it gives proper support to the soft tissues. The cells between the fibres are called *osteocytes* and bone is richly supplied with blood vessels. The structure of bone will be dealt with more fully in Chapter 6.

Haemopoietic tissue is concerned with the formation of the blood cells and is derived from the primitive mesenchyme cells. Blood may be regarded as connective tissue with the plasma forming the matrix which contains the cells. It will be dealt with more fully in Chapter 12.

Muscular Tissue

Muscular tissue is specialized for contraction and is therefore able to produce movement. Wherever there is movement in the body there must be muscular tissue to produce it. Muscle cells are long and thin so that the shortening which occurs during contraction may be as effective as possible. Muscle cells are often called muscle fibres because of their shape. The elastic fibres of connective tissue, if they have been stretched, can recoil and return to their original length, but muscle fibres can shorten themselves without this preliminary stretching. There are three types of muscular tissue; voluntary, involuntary and cardiac.

Voluntary or Striped Muscle

Voluntary muscle forms the flesh of the limbs and trunk, by which the skeleton is moved. It consists of long cells often known as fibres, varying in length from a few millimetres in short muscles to from 30 cm or more in long muscles; each cell contains numerous thread-like fibres called *myofibrils*, which are only from 0·01 to 0·1 mm in width. These myofibrils are regularly striped across in alternate light and dark bands throughout their length and are so arranged that the dark and light parts come next to one another. Each fibril is enveloped in a sheath of connective tissue called the *sarcolemma*. The fibrils are bound together by connective tissue into bundles, each enclosed in a sheath called the *endomysium*; these bundles are in turn bound together and enclosed in a sheath called the *perimysium*, to form ultimately the individual muscles, which also have a sheath of fibrous tissue called the *epimysium*.

The nucleus is at the edge of the cell. This striped muscle is under the control of the will, so that it is known as voluntary muscle. It contracts strongly when stimulated by a nerve fibre, but tires quickly. For strong contraction much energy is required, so voluntary muscle must have a good blood supply to bring oxygen and nutrients to the cells and to carry away waste products. Capillaries run between individual muscle cells to ensure adequate blood supply.

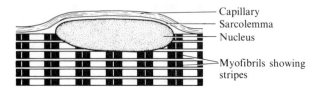

Longitudinal section of muscle fibre

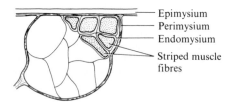

Cross-section of part of muscle

FIG. 22. Sections of voluntary (striped) muscle.

Involuntary or Unstriped Muscle

Involuntary muscle forms the walls of internal organs such as the stomach, bowel, bladder, uterus and blood vessels. It consists of spindle-shaped cells, each of which contains a nucleus. It is also called unstriped muscle. The cells do not show any stripes and have no sheath, but are bound together by connective tissue to form the walls of the various organs. They are not under the control of the will and act without any conscious effort or knowledge. They contract automatically, but are supplied by autonomic nerves which affect their contractions. This type of muscle is designed for slow contraction over a long period and it does not tire easily.

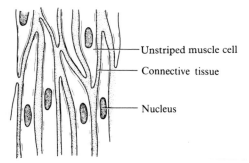

Fig. 23. Section of involuntary (unstriped) muscle.

Cardiac Muscle

Cardiac muscle is both involuntary and irregularly striped. It is found in the heart wall only and is different from any other muscle tissue. It consists of short, cylindrical, branched fibres with centrally placed nuclei. They have no sheath, but are bound together by connective tissue. The cardiac muscle is not under the control of the will, but contracts automatically in a rhythmic manner throughout life, though the rate of these rhythmic contractions is controlled by nerves, which quicken or slow down its action. The fibres branch and join with one another so that impulses can spread from one fibre to another as well as along the length of the muscle.

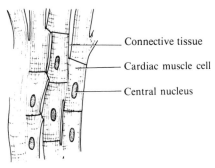

Fig. 24. Section of cardiac muscle.

Nervous Tissue

Nervous tissue is specially designed to receive stimuli from either inside or outside the body, and when stimulated to carry impulses rapidly to other tissues. Nervous tissue consists of nerve cells, called *neurones*, and a special supporting network called *neuroglia*. A nerve cell consists of a large cell body, to which several short processes, called *dendrites*, bring impulses from other cells and tissues. From the cell body there is one long process called the *axon*, which carries impulses away from the cell body.

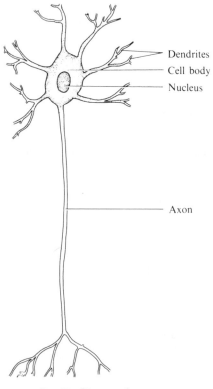

Dendrites
Cell body
Nucleus

Axon

Fig. 25. Diagram of a neurone.

Membranes

The cavities and hollow organs of the body are lined with membranes. These membranes consist of epithelium and secrete lubricating fluids to moisten their smooth, glistening surfaces and prevent friction. Three different types of membrane are found in the body.

Synovial Membrane

Synovial membrane secretes a very thick fluid, which is like white of egg in consistency; hence the name, the prefix 'syn' meaning 'like' (from the Greek) and 'ovum' being Latin for an egg. It is chiefly found lining joint cavities to lubricate the movement of the bones on one another. It is a fibrous membrane covered with epithelium. It is also found over bony prominences and between ligaments and bones or tendons and bones. In these sites it forms small sacs called *bursae*, which act as water cushions, facilitating the movement of one part on another. For example, there are bursae around the shoulder, knee and elbow joints. Synovial membrane is also found forming sheaths for tendons which run a long distance to their insertions. For example, the tendons and muscles in the forearms and legs run across the hand and foot respectively and move the fingers and toes.

Mucous Membrane

Mucous membrane secretes a rather thinner fluid called mucus. This membrane is found lining the food canal from mouth to rectum and the air passages from the nose downwards. Thus the cavities which it lines are connected with the external skin. Mucus-secreting tubular glands are also present where much secretion is produced. These are single or branching tubes, lined with secreting cells.

Serous Membrane

Serous membrane consists of flattened cells through which oozes a small quantity of thin watery fluid called serous fluid; this is similar to the fluid which oozes from a blood clot. Serous membrane is found lining the internal cavities, e.g. the thorax and the abdomen, and covering the organs they contain, providing smooth, glistening, moist surfaces which slide easily over one another when one part moves on, or within, another.

5 Systems and Parts of the Body

As has been mentioned in the previous chapter the human body is an example of a highly developed multicellular organism, its billions of cells especially developed so that each type of tissue carries out specialized functions on behalf of the rest of the body. These tissues are further grouped together to form organs. An *organ* is a group of tissues arranged in a certain way to carry out a specific task, e.g. the stomach, the heart, the kidneys, the spleen. The organs are again grouped together to form systems. A *system* is a group of organs which together carry out one of the essential functions of the body, e.g. the digestive system converts food into simpler substances which can be absorbed and utilized by the body; the respiratory system is involved in bringing oxygen to, and eliminating carbon dioxide from, the blood.

The Systems of the Body

The following systems, grouped together, form the human body.

The *skeletal system* provides a framework which gives support and protection to the soft tissues and allows movement at joints.

The *muscular system* effects movement of the body as a whole. These systems together are sometimes known as the locomotor system.

The *circulatory system* is the transport system of the body; it carries oxygen and nourishment to the tissues and waste products away from them and is essential if there is to be dependence of one tissue on another.

The *respiratory system* allows exchange of gases between the body and the environment.

The *digestive system* is concerned with digestion and absorption of food and the elimination of waste matter.

The *endocrine system* produces hormones which control a variety of functions in the body.

The *urinary system* is the main excretory system of the body.

The *nervous system* creates awareness of the environment and makes it possible for the body to respond to change with the required precision.

The *reproductive system* is responsible for the survival of the species by reproduction of the same kind.

These systems will be dealt with in detail in later chapters, but it is better to start with an overview of the body as a whole, in order to appreciate something of the complexity of the organization of the body and the interdependence of its various parts.

The *skeletal system* consists of many bones forming a rigid supporting framework. Movement is only permitted at the joints, or articulations, where two or more bones meet, but the skeleton has no power to move of itself. The *muscular system* is made up of countless muscles, attached to the bones, which allow them to move. Muscles form the flesh of the body and are responsible for movement from place to place, for the power to reach for and to hold objects, to move the eyes and mouth and to sit, kneel or stand. For this reason the skeletal, articular and muscular systems together are known as the locomotor system.

The *circulatory system* nourishes every part of the body. It includes the blood which carries oxygen, food, waste products and other essential substances from organ to organ, the heart, which acts as a pump to circulate the blood to all parts of the body, and the blood vessels through which the blood travels. The blood vessels are of two types: the arteries which carry blood from the heart to the tissues and the veins which carry blood from the tissues to the heart.

The *respiratory system* consists of air passages leading to the lungs, within which the blood receives a fresh supply of oxygen and given up its excess carbon dioxide.

The *digestive system* is primarily a food canal within which digestive juices act upon ingested food, digestible substances are simplified and absorbed and from which the indigestible residue is eliminated.

The *endocrine glands* consist of specialized cells which are able to extract certain materials from the blood and manufacture from them new substances which exert control over a variety of functions in other parts of the body. The substances which are manufactured by the glands are called hormones and they are secreted directly into the blood, by which they are carried throughout the body to stimulate other organs into activity.

The *urinary system* is the main excretory system of the body. Urine is formed in the kidneys, conveyed by the ureters to the bladder where it is stored until it is convenient for micturition to take place.

The *nervous system* makes it possible for man to be aware of his environment and to respond to changes within that environment. It consists of brain, spinal cord and nerves, some of which carry messages from the tissues to the brain and others which carry messages from the brain to the tissues. Incoming messages are carried by sensory nerves and the brain is able to interpret such messages in the light of experience. Outgoing messages from the brain are carried by motor nerves and result in movement and activity.

The *reproductive system* produces the sex cells, spermatozoa in the male and ova in the female, which together enable the species to continue.

Definition of Terms Used in Anatomy

In order to achieve uniformity of description an *anatomical position* has been chosen and defined. The body is erect, facing the observer, with the arms at the sides and the hands facing forwards.

The following terms are commonly used:

Superior: upper or above
Inferior: lower or below
Anterior or ventral: towards the front
Posterior or dorsal: towards the back
Distal: furthest from the source
Proximal: nearest to the source
External: outer
Internal: inner

The *median or sagittal line* is an imaginary line passing vertically through the mid-line of the body from the crown of the head to the ground between the feet, dividing it into right and left halves. *Lateral* means furthest from the median line; *medial* means nearest to the median line.

A *horizontal* section divides the body into superior and inferior portions.

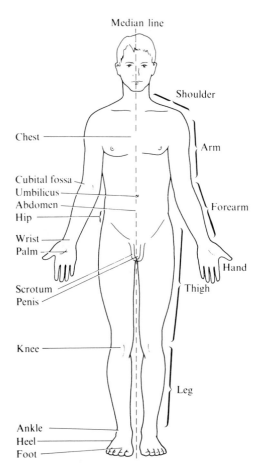

FIG. 26. The anatomical parts of the body (front view).

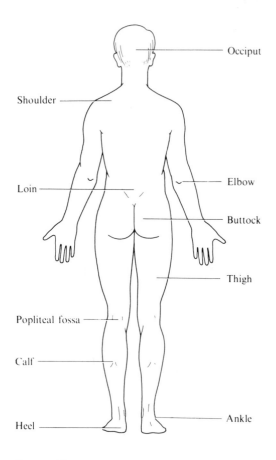

Fig. 27. The anatomical parts of the body (back view).

A *sagittal* section divides the body into right and left portions parallel to the median line.

A *coronal* section divides the body into anterior and posterior portions.

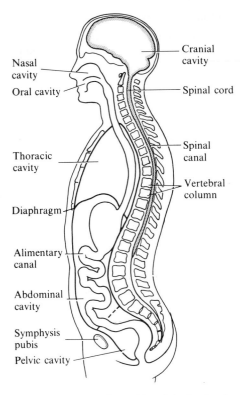

FIG. 28. Diagrammatic section of the cavities of the head and trunk (the unlabelled broken line marks the boundary between the abdomen and pelvic cavity).

Cavities of the Body

The body has two major cavities, each subdivided into lesser cavities.

The *ventral cavity*, within the trunk, is subdivided into:

1. The thorax or thoracic cavity.
2. The abdomen, or abdominal cavity which is continuous with the pelvic cavity.

The *dorsal cavity* is subdivided into:

1. The cranial portion, containing the brain.
2. The spinal portion, containing the spinal cord.

6 Development and Types of Bone

The skeletal system is made up of just over 200 bones, joined together to provide a strong, movable, living framework for the body. The system has four main functions. It supports and protects the surrounding soft tissues and vital organs; it assists in body movement by giving attachment to muscles and providing leverage at joints; it manufactures blood cells in the red bone marrow and it provides storage for mineral salts, particularly phosphorus and calcium.

Cartilage provides the environment in which bone develops. Development is from spindle-shaped cells, called *osteoblasts*, which are found underneath the tough sheath of fibrous tissue which covers the bone called the periosteum, and in cavities within the bone. These cells are able to convert the soluble salts of calcium and magnesium in the blood, such as calcium chloride, into insoluble calcium salts, chiefly calcium phosphate. A second set of cells is also present which can change the insoluble calcium phosphate into soluble calcium salts which are carried away by the blood. These cells are capable of causing the absorption of any unwanted bone and are known as *osteoclasts*. Both types of cell are active during the growth period, osteoblasts producing bone and osteoclasts removing it to maintain the form and proportion of the bone. For example, osteoblasts build up bone on the surface of a hollow bone, while osteoclasts absorb bone on the inner side to enlarge the cavity and prevent the bone from becoming too heavy.

Ossification

There are two types of ossification. *Intramembranous ossification* is a process by which dense connective tissue is replaced by deposits of calcium salts, forming bone. The bones of the skull form by this process.

Most bones form by *intracartilaginous ossification* in which cartilaginous structures are replaced with bone.

Bone Growth and Repair

Plentiful supplies of calcium and phosphorus are essential in the diet of the pregnant and nursing mother, the growing child and where bone repair is taking place after injury or disease. Calcium is present in milk and eggs and in green vegetables and phosphorus in meat, fish and the yolks of eggs. Vitamin D is necessary so that calcium and phosphorus may be absorbed from the intestine and fully utilized within the body; lack of vitamin D causes rickets in children and osteomalacia in adults. Both these conditions cause soft bones, which bend under the weight of the body and various deformities of the weight-bearing bones may result. Vitamin D is found in animal fat and fish oils and in margarine which has been artificially irradiated.

The human body can also manufacture the vitamin. The action of the ultra-violet rays of the sun converts *ergosterol* in the skin into vitamin D.

Vitamin C is important in bone growth as it plays a part in laying down *collagen*, which is the main constituent of connective tissue. It is found in fresh fruit, particularly citrus fruit, blackcurrants, green vegetables, tomatoes and potatoes.

Growth and development of bone is also affected by both *exercise* and *rest*. Exercise causes an increased blood supply to the muscles and therefore to the underlying bones; because the blood carries necessary building materials exercise results in increased growth and recognition of this fact is responsible for increased attention to physical exercise in schools. The pull of developing muscles is an important factor in determining the shape of the bones and posture may also affect shape by altering stresses on bones. The importance of rest is interesting. The bones of a child are comparatively elastic so that at the end of a day of standing and running about there is an appreciable loss of height. During rest the bones recover their full length, so a mid-day rest and a long night will affect growth.

A further factor controlling growth is the secretions of certain ductless glands. This will be mentioned again in Chapter 20.

Types of Bone Tissue

Bone tissue is of two types, compact and spongy.

Compact bone appears to be solid but when it is examined under a microscope it is found to consist of *Haversian systems*. A Haversian system consists of:

1. A *central canal*, called a Haversian canal which contains blood vessels, nerves and lymphatics.
2. Plates of bone, called *lamellae*, arranged round the central canal.
3. Between the lamellae are spaces called *lacunae* which contain bone cells, called osteocytes, and lymph.
4. Fine channels, called *canaliculi*, between the lacunae and the central canal carry lymph which brings nutrients and oxygen to the osteocytes.

Between the Haversian systems there are tiny circular plates of bone called *interstitial lamellae*.

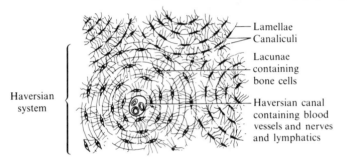

FIG. 29. The structure of compact bone.

Spongy bone is hard like all bone but has a spongy appearance to the naked eye. When examined under a microscope the Haversian canals are seen to be much larger and there are fewer lamellae. The spaces in spongy bone are filled with *red bone marrow* which consists of fat and blood cells and in which red blood cells are made.

Types of Bones

There are three types of bones:

1. Long bones
2. Flat bones
3. Irregular bones

Long Bones

A long bone consists of a shaft and two extremities. The shaft has an outer layer of compact bone surrounding a central cavity called the *medullary canal* which contains yellow bone marrow. Yellow bone marrow, like red bone marrow, consists of fat and blood cells, but it is not so rich in blood supply or in red blood cells. The extremities consist of a mass of spongy bone containing red bone marrow covered by a thin layer of compact bone. The bone is covered by a tough sheath of fibrous tissue called the *periosteum*; this is richly supplied with blood vessels which pass into the bone to nourish it.

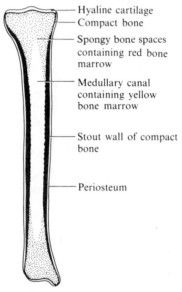

Hyaline cartilage
Compact bone
Spongy bone spaces containing red bone marrow
Medullary canal containing yellow bone marrow
Stout wall of compact bone
Periosteum

FIG. 30. The structure of a long bone.

Three distinct sets of blood vessels supply the long bones:

1. Innumerable tiny arteries run into the compact bone to supply Haversian canals and systems.

2. Many larger arteries pierce the compact bone to supply the spongy bone and red bone marrow. The openings by which these vessels enter the bone can be seen quite easily by the naked eye.

3. One or two large arteries supply the medullary canal. These are known as nutrient arteries and they pass through a large

opening called the nutrient foramen, which runs obliquely through the shaft to the medullary canal.

These three sets of vessels are linked by fine branch arteries within the bone.

The periosteum nourishes the underlying bone through its blood vessels. If it is torn off the underlying bone will die, but if the bone is destroyed by disease and the periosteum remains healthy it is able to build up new bone. The periosteum is also responsible for growth in the thickness of the bone through the action of osteoblasts, adjacent to the bone surface, which are able to lay down fresh bone. The periosteum is protective in function and also gives attachment to tendons of muscles. It is not present on the joint surfaces of a bone where it is replaced by hyaline cartilage, called articular cartilage, which provides a smooth surface so that joint movements can occur without friction.

Development of Long Bones. Long bones develop from three centres of ossification, one in the shaft and one or more at either extremity. The centre of ossification in the shaft is known as the

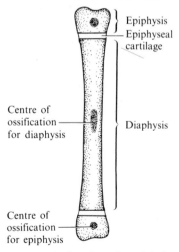

Fig. 31. The development of a long bone (growth in length occurs at the epiphyses).

diaphysis. Those at the extremities are called *epiphyses* and begin to develop after birth. From these centres ossification gradually spreads throughout the extremity and is well developed by 12 years of age, though there is still a line of cartilage left between the shaft and the extremity.

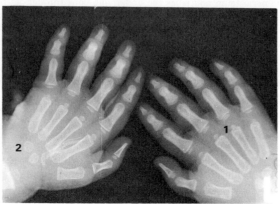

FIG. 32. X-ray of the hand of a 1-year-old infant. The small long bones of the fingers and the thumb have their shafts well formed, but note the gaps (1) between them and the knuckles, where the epiphyses are only beginning to form centres of ossification. Note also that only two of the eight bones in the wrist (2) have begun to ossify.

The line of cartilage between epiphysis and diaphysis is known as the *epiphyseal cartilage* and it is from this cartilage that growth in length occurs. It is the shaft that grows longer and fresh bone is constantly produced at either end of it by the epiphyseal cartilage. When full growth has been attained, between the ages of 18 and 25 years, the line of cartilage turns to bone and no further growth can occur. The timing of this event varies from one bone to another and from one end of a bone to the opposite end of the same bone. For example, the lower epiphysis of the humerus unites at about the 18th year, but the upper epiphysis does not join the shaft until about two years later.

Flat Bones

Flat bones consist of two stout layers of compact bone joined by a layer of spongy bone. These bones are also covered by periosteum,

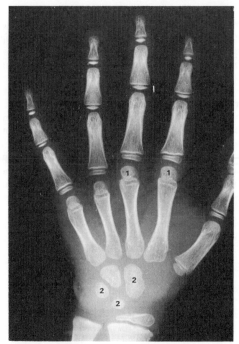

Fig. 33. X-ray of the hand of a 6-year-old child. Note (1) the development of the separate extremities, i.e. the epiphyses, of the long bones of the hand; and (2) the ossification that has occurred in the wrist or carpal bones. Only certain bones are marked for comparison.

from which two sets of blood vessels pass into the bone to supply the compact and spongy bone respectively. They are found in the head, trunk, shoulder girdle and pelvic girdle, where they give protection to the delicate organs underlying them and provide attachment for the powerful muscles required to control the freely movable shoulder

Compact bone — Spongy bone

Fig. 34. The structure of a flat bone.

and hip joints. The layers of compact bone in the flat bones of the skull are referred to as the *outer* and *inner tables* and the spongy bone is called the *diploë*. The outer table is thick and strong so that it is not readily broken. In some areas the spongy bone is absorbed leaving air-filled cavities between the two tables which are called *sinuses*.

Irregular Bones

These consist of a mass of spongy bone covered by a thin layer of compact bone. They are covered with periosteum, except on the articular surfaces, and as with flat bones this protective covering provides two sets of blood vessels to supply both compact and spongy bone. They are found in the spinal column and the middle ear and also at the ankle and wrist, although these latter are sometimes called short bones.

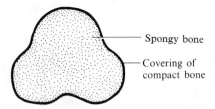

Spongy bone

Covering of compact bone

FIG. 35. The structure of an irregular bone.

Surface Irregularities

The surfaces of all bones are very irregular and show numbers of *depressions* and *projections*. These may be divided according to function into:

1. Articular, which enter into the formation of joints and are smooth.
2. Non-articular, which give attachment to muscles or ligaments and are rough.

Articular projections are termed:

1. A head, when round like a sphere or disc.
2. A condyle, when rounded but oval in outline like the typical knuckle bone.

Articular depressions are termed sockets or fossae.

Non-articular projections have many terms applied to them according to their nature.

1. Process—a rough projection for muscle attachment
2. Spine—a pointed rough projection
3. Tuberosity—a broad rough projection
4. Trochanter—a broad rough projection
5. Tubercle—a small tuberosity
6. Crest—a long rough narrow projecting surface

All these rough projections give attachment to muscles; the stronger the muscle and the more it is used, the larger and rougher the projection, providing a greater surface for muscle attachment. In a paralysed limb the processes either fail to develop, or atrophy, according to age.

Non-articular depressions are termed:

1. Fossa—a notch in the bone
2. Groove—a long, narrow depression

Other terms applied to bones are:

1. Foramen—an opening in the bone
2. Sinus—a hollow cavity in the bone

7 Bones of the Head and Trunk

The student is advised that it is not possible to study adequately the bones of the skeleton without access to an entire skeleton and to disarticulated bones which can be handled and examined closely.

The skeleton can be divided into:

1. The bones of the head
2. The bones of the trunk
3. The bones of the upper limb and shoulder girdle
4. The bones of the lower limb and pelvic girdle

The bones of the head and trunk are also known as the *axial* skeleton, the main support of the body, while the bones of the extremities may be known as the *appendicular* skeleton.

The Bones of the Head

For purposes of description the skull may be divided into:

1. The bones of the cranium
2. The bones of the face

The Bones of the Cranium

The cranium is a box-like cavity which contains and protects the brain. It has a dome-shaped roof called the *calvaria* or skull cap and its floor is known as the *base of the skull*. The cranium consists of fifteen bones:

One frontal bone
Two parietal bones
One occipital bone
Two temporal bones
One ethmoid bone
One sphenoid bone

Two inferior nasal conchae
Two lacrimal bones
Two nasal bones
One vomer

The *frontal bone* is a large flat bone forming the forehead and most of the roof of the orbit. There are rounded prominences, called

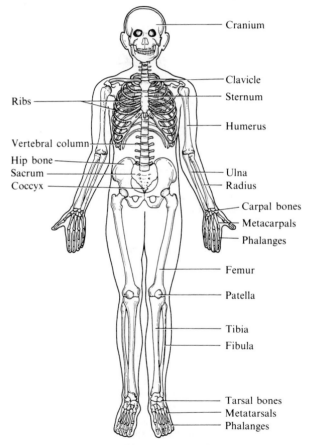

FIG. 36. The skeleton (male).

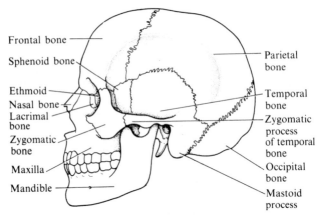

Frontal bone

Sphenoid bone

Ethmoid

Nasal bone

Lacrimal bone

Zygomatic bone

Maxilla

Mandible

Parietal bone

Temporal bone

Zygomatic process of temporal bone

Occipital bone

Mastoid process

FIG. 37. The bones of the head.

the *frontal tuberosities*, one on each side of the midline, which vary in size from one individual to another and which together form the forehead. The bone contains two irregular cavities called the *frontal sinuses* which lie one over each orbit and which open into the nasal cavity. The sinuses contain air and are lined with mucous membrane which is continuous with the mucous membrane lining the respiratory tract. They add resonance to the voice and they serve to lighten the skull, but the mucous membrane may become infected, causing a condition known as *sinusitis*.

The *parietal bones* form the sides and roof of the cranium; they articulate with the frontal bone, the occipital bone and with each other to form the *sutures* or joints of the cranium (see Chapter 9). On the internal surface are small grooves to carry the blood vessels supplying the brain and the impression of the folds or convolutions of the surface of the brain can be seen. At birth there are membranous gaps in the skull at the angles of the parietal bone which are called *fontanelles* (see Chapter 9).

The *occipital bone* forms the back of the skull. It carries a marked prominence, the *external occipital protuberance* which gives attachment to muscles. Below this there is a large oval opening, known as the *foramen magnum*, through which the cranial cavity communicates with the vertebral canal. On either side of the foramen are two smooth oval processes called the *occipital condyles* for articulation

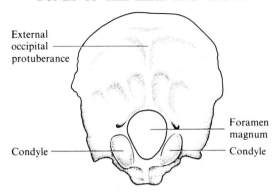

External occipital protuberance

Foramen magnum

Condyle

Condyle

Fig. 38. The occipital bone (from below).

with the first cervical vertebra. This joint allows the nodding movement of the head.

The *temporal bones* are situated at the sides and base of the skull. Each consists of four parts:

1. The *squamous part* forms the anterior and upper part of the bone and is thin and flat. A long arched process, called the *zygoma* or zygomatic process projects forward from the lower portion.

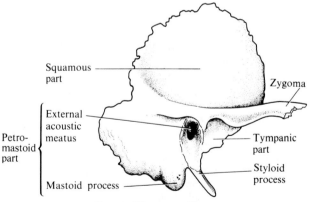

Squamous part

Zygoma

Petro-mastoid part

External acoustic meatus

Tympanic part

Styloid process

Mastoid process

Fig. 39. The temporal bone.

2. The *petromastoid part* forms the posterior portion of the bone and can be divided into two parts:

    a. The *mastoid portion* which continues into a conical projection called the *mastoid process*, containing air cells.

    b. The *petrous portion*, between the occipital bone and the sphenoid, which contains the structures forming the internal ear.

3. The *tympanic part* is a curved plate lying below the squamous part and in front of the mastoid process. It contains the external acoustic meatus.

4. The *styloid process* projects downwards and forwards from the underneath of the bone.

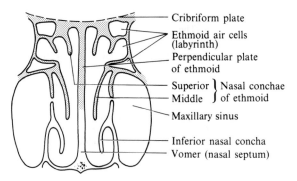

Cribriform plate

Ethmoid air cells (labyrinth)

Perpendicular plate of ethmoid

Superior } Nasal conchae
Middle } of ethmoid

Maxillary sinus

Inferior nasal concha

Vomer (nasal septum)

Fig. 40. Diagram of section of ethmoid (shown in colour).

The *ethmoid bone* is very light and irregular in shape and consists of three parts:

1. The *cribriform plate*, a small horizontal plate which is perforated with many fine openings, or foramina, for the passage of the *olfactory nerves*, which transmit the sense of smell.

2. The *perpendicular plate*, which descends from the cribriform plate, and forms the upper part of the nasal septum, which divides the nasal cavity into two.

3. Two *labyrinths*, each consisting of a number of thin-walled *ethmoidal air cells* which communicate with the nasal cavity and may become infected from it. Two thin plates of bone called the *superior and middle nasal conchae*, jut out into the nasal cavities from the spongy labyrinths.

The *sphenoid bone* is situated at the base of the skull, in front of the temporal bones. It is shaped rather like a bat with outstretched wings. The *body* contains two large air sinuses, which communicate with the nasal cavity, and a deep depression, known as the *hypophyseal fossa*, which contains the *hypophysis cerebri* or pituitary gland. The *greater and lesser wings* are perforated by many openings for the passage of nerves and blood vessels.

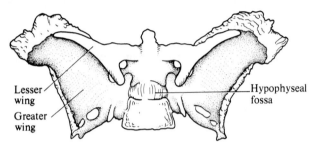

Lesser wing
Greater wing
Hypophyseal fossa

Fig. 41. The sphenoid bone.

The *inferior nasal conchae* are curved plates of bone which lie in the walls of the nasal cavity, below the superior and middle nasal conchae of the ethmoid bone (see Fig. 40).

The *lacrimal bones* are the smallest and most fragile of the cranial bones and form part of the walls of the orbits. Each is grooved to contain the *lacrimal sac* and *nasolacrimal duct*, by which the *lacrimal fluid*, or tears, which washes constantly over the surface of the eye, is carried into the nasal cavity.

The *nasal bones* are two small oblong bones which together form the bridge of the nose.

The *vomer* is a flat bone which forms the lower part of the septum of the nose.

The Bones of the Face

The bones of the face are:

> The maxillae
> The mandible
> Two zygomatic bones
> Two palatine bones
> The hyoid bone

The *maxillae* are the largest bones of the face (excepting the mandible) and by their union in the midline form the whole of the upper jaw. They carry the upper teeth embedded in a ridge of bone called the *alveolar process*. The *palatine process* of the maxilla is a horizontal projection which forms a considerable part of the floor of the nasal cavity and the roof of the mouth. The *maxillary sinus* is an air-filled cavity within the body of the bone which communicates with the nasal cavity and which may become infected following nasal infection.

The *mandible* is an irregular bone and is the only movable bone in the head. It forms the lower jaw and carries the lower teeth embedded in the *alveolar part*. The *rami* are vertical projections which

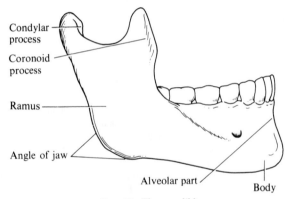

Condylar process
Coronoid process
Ramus
Angle of jaw
Alveolar part
Body

FIG. 42. The mandible.

carry a *condylar process* which articulates with the temporal bone and a *coronoid process* which provides attachment for muscle. The upright and horizontal portions meet to form the angles of the jaw.

The *zygomatic bones* are irregular bones which form the prominence of the cheek and part of the walls of the orbit. The *temporal process* articulates with the zygomatic process of the temporal bone to form the *zygomatic arch*.

The *palatine bones* are irregular bones which form part of the hard palate, the lateral wall of the nasal cavity and the floor of the orbit.

The *hyoid bone* is U-shaped and lies at the base of the tongue, to which it gives attachment. It does not articulate with any other bone

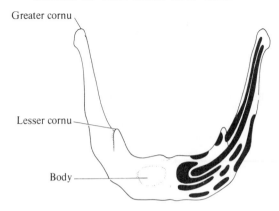

Greater cornu

Lesser cornu

Body

FIG. 43. The hyoid bone.

but is connected by ligaments to the styloid processes of the temporal bone (see Fig. 39).

The Bones of the Trunk

The bones of the trunk are:

> The sternum
> The ribs
> The vertebral column

The Sternum

The sternum is a long flat bone which runs down the front of the thorax close under the skin. Its upper end supports the *clavicles* and it also articulates with the first seven pairs of ribs. The bone is in three parts:

1. The *manubrium* is triangular in shape and its lower border is covered with a thin layer of cartilage for articulation with the upper end of the body.

2. The *body* is longer and narrower than the manubrium. Where it joins the manubrium there is a small notch which accommodates the cartilage of the second rib.

3. The *xiphoid process* is small and variable in shape and may not become completely ossified.

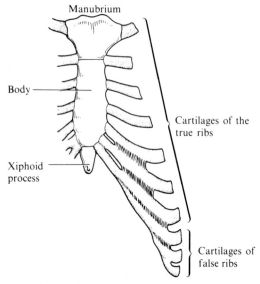

FIG. 44. The sternum and costal cartilages.

The Ribs

The ribs are arched bones which are connected behind with the vertebral column. There are usually twelve pairs, the first seven of which are attached to the sternum by the *costal cartilages* and are known as the true ribs. The remaining five are called false ribs. Of these the upper three are joined to the cartilage of the rib immediately above and the lower two are free at their anterior ends and are known as floating ribs.

The ribs form the curved walls of the thorax, sloping downwards towards the front. They increase in size from above downwards so that the thoracic cavity is roughly cone-shaped.

Each rib is curved and the under surface is grooved for the passage of intercostal arteries, veins and nerves. The *vertebral end* possesses a head, a neck and a tubercle. The head has two smooth facets which articulate with the bodies of the corresponding vertebrae.

The *tubercle* has a small oval facet for articulation with the transverse process of the corresponding vertebra.

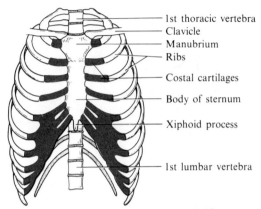

1st thoracic vertebra
Clavicle
Manubrium
Ribs

Costal cartilages

Body of sternum

Xiphoid process

1st lumbar vertebra

THORACIC CAGE FROM THE FRONT

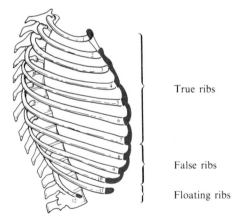

True ribs

False ribs

Floating ribs

THORACIC CAGE FROM THE SIDE

FIG. 45. The thoracic cage.

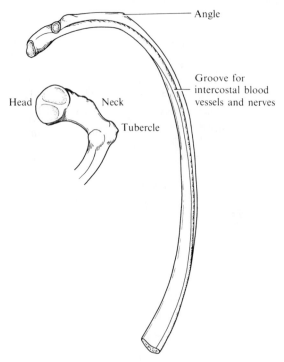

Angle

Groove for
intercostal blood
vessels and nerves

Head   Neck

Tubercle

FIG. 46. A typical rib.

The Vertebral Column

The vertebral column consists of a number of irregular bones
called the *vertebrae* which are firmly connected to one another but
are capable of a limited amount of movement on one another. The
column provides a central axis and also protects the spinal cord,
which it surrounds. Each vertebra consists of a cylindrical *body*
lying to the front, and a *vertebral arch* projecting backwards and
enclosing a space called the *vertebral foramen* through which the
spinal cord passes. The arch has a *spinous process* directed backwards and downwards and two *transverse processes* which project
laterally. These are for the attachment of muscles and ligaments. On

the under surface of the arch is a notch for the passage of spinal nerves and vessels. Each has four articular processes, two above and two below, which meet the corresponding processes of adjoining vertebrae.

The wide parts of the arch carrying the spinous process are known as the *laminae*; this is the part removed in the operation of laminectomy to relieve pressure on the spinal cord caused by disease or following injury.

The bodies of adjoining vertebrae are firmly connected to one another by discs of fibrocartilage called *intervertebral discs*. Each disc has an outer ring of fibrocartilage and an inner core called the *nucleus pulposus*. It is possible for the outer ring to become weakened and for the nucleus pulposus to irritate an adjoining nerve root causing pain.

The vertebrae are divided into five groups:

1. Seven cervical vertebrae
2. Twelve thoracic vertebrae
3. Five lumbar vertebrae
4. Five sacral vertebrae
5. Four coccygeal vertebrae

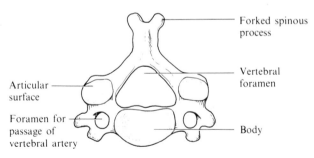

FIG. 47. A typical cervical vertebra.

The seven *cervical vertebrae* are the smallest and can be easily identified because their transverse processes are perforated by foramina for the passage of the vertebral arteries. The spinous process is forked and gives attachment to muscles and ligaments.

The first cervical vertebra is called the *atlas*. It has no body or spine but consists of a ring of bone with two facets which articulate

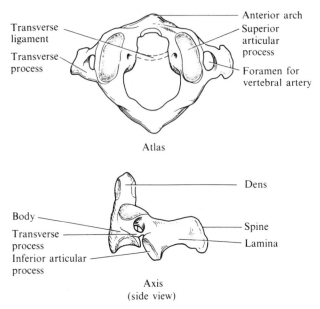

Atlas

Axis
(side view)

Fig. 48. The first and second cervical vertebrae.

with the occipital bone. A ligament called the transverse ligament
divides the ring of the atlas into two.

The *axis*, or second cervical vertebra, carries a strong tooth-like
process called the *dens* (or odontoid process) which projects upwards
from the body and provides a pivot on which the atlas, and therefore the skull, rotate, allowing a turning movement of the head. The
dens is retained in position by the transverse ligament of the atlas,
behind which lies the spinal cord.

The seventh cervical vertebra is distinguished by a very long
spinous process which can be seen and felt through the skin at the
base of the neck and which is not forked.

The twelve *thoracic vertebrae* are larger than the cervical and
they increase in size from above downwards. The *body* is roughly
heart-shaped and has two distinguishing features:

1. Additional facets on each side to articulate with the head and
tubercle of the corresponding rib.

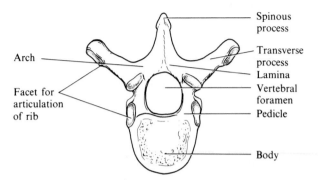

FIG. 49. A typical thoracic vertebra (from above).

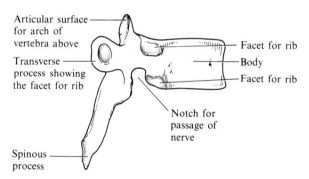

FIG. 50: A typical thoracic vertebra (side view).

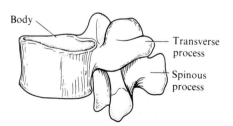

FIG. 51. A lumbar vertebra.

2. Long pointed spinous processes which project downwards and backwards.

The heads of the ribs lie between the vertebrae and articulate with one facet on the vertebra above and one on the vertebra below.

The five *lumbar vertebrae* are the largest vertebrae and have no facets for articulation with the ribs. The spinous processes are large and strong and give attachment to the muscles.

The five *sacral vertebrae* are fused together to form a large bone, the sacrum, which is triangular in shape and forms a wedge between the two hip bones with which it articulates. The pelvic surface of the bone is concave and the anterior projection at the upper end is known as the *sacral promontory* (see Fig. 53). The vertebral foramen found in the other vertebrae is here called the *sacral canal*, and from it four openings allow for the passage of nerve roots.

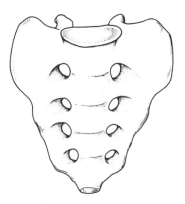

Fig. 52. The sacrum.

The *coccyx* is a small triangular bone which consists of four vertebrae fused together. It articulates with the sacrum and the joint allows slight movement backwards and forwards which increases the size of the pelvic outlet during childbirth.

The vertebral column is the main support of the head and trunk as well as giving protection to the spinal cord. When viewed from the side it has four curves; the thoracic and pelvic curves are termed primary curves as they are present during fetal life. The cervical and lumbar curves are secondary as they appear or are accentuated

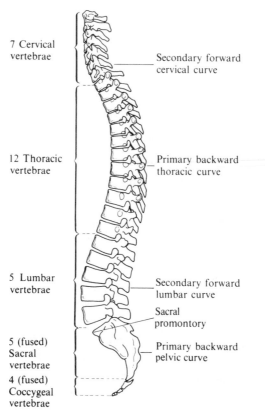

7 Cervical vertebrae

Secondary forward cervical curve

12 Thoracic vertebrae

Primary backward thoracic curve

5 Lumbar vertebrae

Secondary forward lumbar curve

Sacral promontory

5 (fused) Sacral vertebrae

Primary backward pelvic curve

4 (fused) Coccygeal vertebrae

FIG. 53. The curves of the spine.

when the child begins to hold up its head and sit up (cervical) and when it begins to stand and walk (lumbar).

There is only limited movement between any two adjoining vertebrae but there is considerable movement in the vertebral column as a whole. The intervertebral discs cushion any jarring which may occur; as for example jumping from a height and landing on the feet. The curves of the spine enable it to bend without breaking, but a blow on the column is more likely to cause a fracture or a dislocation because the vertebrae are so firmly united to one another.

8 Bones of the Limbs

Bones of the Upper Limb

The bones of the upper limb are:

The scapula ⎫
The clavicle ⎬ forming the shoulder girdle
The humerus
The radius ⎫
The ulna ⎬ forming the forearm
Eight carpal bones
Five metacarpal bones
Fourteen phalanges

The *scapula* is a triangular flat bone; it lies over the ribs at the back of the thorax, but does not articulate with them. It is held in

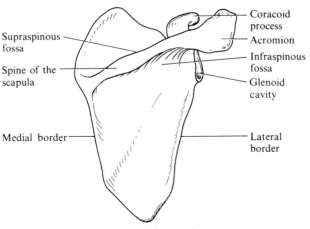

Supraspinous fossa

Spine of the scapula

Medial border

Coracoid process

Acromion

Infraspinous fossa

Glenoid cavity

Lateral border

Fig. 54. The scapula.

place by muscles which attach it to the ribs and vertebral column. This arrangement gives great freedom of movement to the shoulder girdle, making it possible to reach widely both forwards and backwards and to either side of the body. It is not often broken by falls as it is embedded in muscle. It has three borders and three angles; the lowest is spoken of as the angle because it is the sharpest and is easily felt. The front surface is concave to fit over the ribs; the posterior surface is convex and carries a projecting ridge of bone known as the *spine of the scapula*, which gives attachment to muscles and forms two depressions or fossae, one above and one below it.

The outer angle carries a shallow socket known as the *glenoid cavity*, which receives the head of the humerus to form the shoulder joint. Above this two processes project:

1. The *acromion*, the larger of the two, which overlaps the socket and articulates with the clavicle to form the shoulder girdle.

2. The *coracoid process*, which juts forward and is like a hook.

Both of these can easily be felt. They serve to give attachment to muscles and also help to keep the head of the humerus in place, preventing upward dislocation.

Acromial extremity

Sternal extremity

FIG. 55. The clavicle.

The *clavicle* is a long bone, roughly S-shaped. It articulates with the sternum at its inner or sternal extremity and with the scapula at its outer or acromial extremity. The two extremities are easily distinguishable from one another. The inner extremity is roughly like a pyramid in shape, while the outer is flatter and very similar in shape and form to the acromion process of the scapula, with which it articulates. The bone lies close under the skin and is easily felt along its whole course; starting from the sternal extremity it curves first forwards and then backwards. It keeps the scapula in position and when it is broken the shoulder drops forwards and downwards. It is the only bony link between the bones of the upper limb and the axial skeleton, as the scapula does not articulate with either the ribs or vertebral column. It is a bone which is not found in the skeleton of many four-footed animals, as it only becomes necessary to fix the scapula when the limb is moved outwards from the trunk. The bone is easily broken by falls on the shoulder, as it is compressed between the sternum and the point of impact; it is, in fact, better that it should break than that there should be an injury at the root of the neck, where there are many important structures, or about the actual shoulder joint, where injury would be likely to limit subsequent movement.

The *humerus* is the largest and longest bone of the upper limb. The upper extremity has a hemispherical *head*, covered with hyaline cartilage, which articulates with the glenoid cavity of the scapula, forming the shoulder joint. The *anatomical neck* forms a slight constriction adjoining the head and the *greater* and *lesser tubercles* lie below the neck and give attachment to muscles. Between the tubercles a deep groove accommodates one of the tendons of the biceps muscle. The shaft of the humerus has many roughened

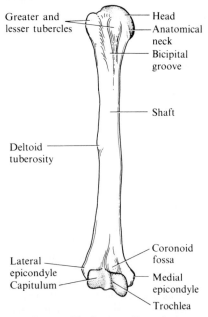

Greater and lesser tubercles

Deltoid tuberosity

Lateral epicondyle
Capitulum

Head
Anatomical neck
Bicipital groove

Shaft

Coronoid fossa
Medial epicondyle
Trochlea

FIG. 56. The humerus (front view).

surfaces for the attachment of muscles, the most marked being the *deltoid tuberosity* on the outer side, which gives insertion to the deltoid muscle. A groove running obliquely round the shaft carries the radial nerve, one of the three main nerves of the upper limb.

The lower end of the humerus is divided into articular and non-articular portions. The *articular portion*, together with the radius and the ulna, forms the elbow joint. It is divided by a shallow groove into the *capitulum*, a rounded projection which articulates with the radius, and the *trochlea* which is shaped rather like a pulley and which articulates with the ulna. The *non-articular portion* has two *epicondyles* which give attachment to muscles. There are also two deep hollows; the posterior one is called the *olecranon fossa* because it accommodates the olecranon process of the ulna when the elbow is extended and the *coronoid fossa*, which is on the anterior surface, provides for the coronoid process of the ulna when the elbow is flexed.

The *radius* is the outer bone of the forearm. The upper end is smaller and has a disc-shaped *head* with a hollowed upper surface to articulate with the capitulum of the humerus. The head also articulates with the ulna. The *neck* of the radius is the constricted portion below the head and on the ulna side there is a projection called the *radial tuberosity* which gives insertion to the biceps muscle. The *shaft* of the bone has a sharp ridge facing the ulna and from it a sheet of fibrous tisue called the *interosseous membrane* runs to the ulna, connecting the two bones. The lower end of the radius is the wider part and takes part in the formation of the wrist joint; it also has a projection called the *styloid process* which can be felt at the base of the thumb.

The *ulna* lies on the inner side of the forearm. The upper end is shaped like a hook and has two large projections; the *olecranon* fits into the olecranon fossa of the humerus when the arm is straight and its upper border forms the point of the elbow. It provides insertion for the tendon of the triceps muscle. The *coronoid process* is smaller and projects forwards. These two processes help in the formation of the *trochlear notch* which articulates with the trochlea of the humerus. The *radial notch* is a depression on the upper part of the coronoid process which articulates with the head of the radius and allows the turning movement of the hand. When rotation of the hand takes place the lower end of the radius is carried round the lower end of the ulna so that the shafts of the two bones cross each other in the middle of the forearm.

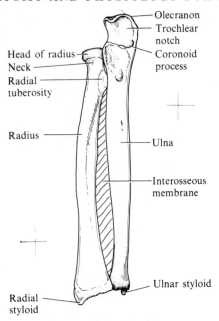

Olecranon
Trochlear notch
Coronoid process
Head of radius
Neck
Radial tuberosity
Radius
Ulna
Interosseous membrane
Ulnar styloid
Radial styloid

FIG. 57. The radius and ulna.

The *shaft* of the ulna, like that of the radius, carries a sharp ridge for the attachment of the interosseous membrane which lies between the two bones.

The lower end of the ulna has a rounded part, known as the *head*, and a projection called the *styloid process*. The head articulates with the ulnar notch of the radius. The styloid process gives attachment to a ligament of the wrist joint; it may be felt below the skin of the wrist and is sometimes quite prominent.

The *carpal bones* (or carpus) comprise eight bones which are arranged in two rows of four. The bones of the upper row are named the *scaphoid*, the *lunate*, the *triquetral* and the *pisiform*. The first three articulate with the radius. The bones of the lower row are called the *trapezium*, the *trapezoid*, the *capitate* and the *hamate*. The surface of the carpus which forms the palm has a deep cavity, the carpal groove, which has a fibrous band across it and which

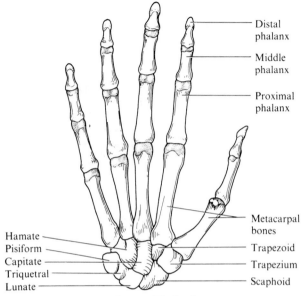

Distal phalanx

Middle phalanx

Proximal phalanx

Metacarpal bones

Trapezoid

Trapezium

Scaphoid

Hamate
Pisiform
Capitate
Triquetral
Lunate

FIG. 58. The hand.

accommodates the median nerve and some of the tendons to the hand. This is known as the *carpal tunnel*.

The *metacarpal bones* (or metacarpus) are five miniature long bones running across the palm of the hand. The *bases* of the bones articulate with the lower bones of the carpus; the heads articulate with the phalanges. The first metacarpal, which articulates with the two phalanges which form the thumb, can be moved more freely than the other four and can be opposed to each finger in turn, thereby increasing the power of the grip.

The *phalanges* are also miniature long bones, three for each finger and two for the thumb.

Bones of the Lower Limb

The bones of the lower limb are:

> The hip bone, which forms part of the pelvis
> The femur

The patella
The tibia ⎫
The fibula ⎭ forming the leg
Seven tarsal bones
Five metatarsal bones
Fourteen phalanges

The *hip bone* is a large, irregularly shaped bone which articulates in front with the corresponding bone of the opposite side. Each bone consists of three parts, the *ilium*, the *ischium* and the *pubis* which are united at the deep cavity on the outer aspect of the bone called the *acetabulum*. Full ossification is not completed until between the ages of fifteen and twenty-five and before the bones are united by cartilage.

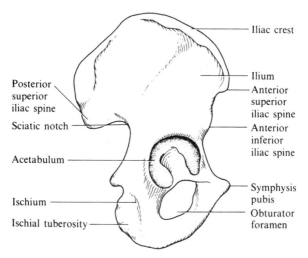

FIG. 59. The hip bone.

The *ilium* includes the upper part of the acetabulum and the expanded flattened area of bone above it which is called the *iliac crest* and which gives attachment to the lateral muscles of the abdominal wall. The crest projects a little beyond the lower part of the bone and ends in front at the *anterior superior iliac spine* which is easily felt at the lateral end of the fold of the groin. The *posterior superior iliac spine* underlies a small dimple which is readily seen on the lower part of the back. The ilium also carries the *anterior* and *posterior inferior iliac spines* which provide attachment for powerful muscles. At the back there is an articular surface for articulation with the sacrum and below this there is a notch called the *greater sciatic notch* for the passage of part of the sciatic nerve.

The *ischium* forms the lower posterior part of the hip bone. It carries the *ischial tuberosity* which gives attachment to muscles and supports the body in the sitting position.

The *pubis* forms the anterior part of the hip bone and meets the pubis of the opposite side in a cartilaginous joint called the *symphysis pubis*. The pubis consists of a *body* which enters into the symphysis and two branches, one running up to join the ilium and the other down to join the ischium. Between these branches and the ischium is a large opening called the *obturator foramen* which is filled with a sheet of fibrous tissue.

The *acetabulum* is the deep socket in the centre of the lower part of the hip bone into which the head of the femur fits.

The *pelvis* is a bony ring composed of the two hip bones and the sacrum and coccyx behind. It is divided into the greater (false) and the lesser (true) pelvis by the *linea terminalis* and the promontory of the sacrum. The *greater pelvis* is the upper expanded portion bounded on each side by the ilium and at the back by the base of the sacrum. The *lesser pelvis* consists of a short curved canal deeper at the back than the front. The female pelvis is shorter than the male

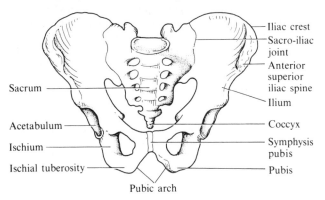

FIG. 60. The male pelvis.

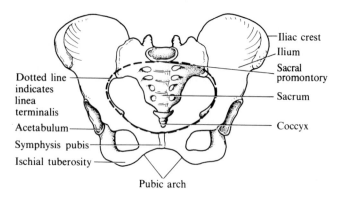

Iliac crest
Ilium
Sacral promontory
Dotted line indicates linea terminalis
Sacrum
Acetabulum
Coccyx
Symphysis pubis
Ischial tuberosity
Pubic arch

FIG. 61. The female pelvis.

and is also wider. The brim of the female pelvis is larger and more nearly circular than that of the male which is more typically heart-shaped.

The *femur* is the longest and strongest bone in the body. The upper end of the bone has a hemispherical *head* which articulates with the acetabulum of the hip; near its centre is a small depression called the *fovea* (see Fig. 70), which gives attachment to the ligament of the head of the femur; this ligament runs to the base of the acetabulum. The *neck of the femur* projects at a marked angle from the shaft and enables free movement of the hip joint. Where the neck joins the shaft there are two processes, the *greater* and *lesser trochanters* which give attachment to muscles. The greater trochanter is on the outer side and can be felt under the skin. The *shaft* of the femur is thinnest in the middle and widens considerably at the lower end. The posterior border is formed by a roughened ridge called the *linea aspera* which gives attachment to muscles.

The lower end of the femur is widely expanded so that there is a good area for the transmission of body weight to the tibia. It has two prominent condyles which articulate with the tibia; they are separated at the back by a deep gap called the *intercondylar fossa* and united in front by a smooth surface which articulates with the patella. Above the condyles at the back of the bone is the *popliteal surface* which forms part of the *popliteal fossa*, containing blood vessels and nerves.

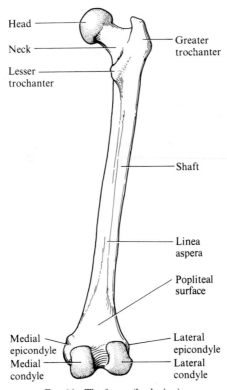

Head

Neck

Lesser trochanter

Greater trochanter

Shaft

Linea aspera

Popliteal surface

Medial epicondyle
Medial condyle

Lateral epicondyle
Lateral condyle

FIG. 62. The femur (back view).

The *patella* is situated in front of the knee joint in the tendon of the quadriceps muscle, which straightens the knee. Bones which develop in tendons in this way are called *sesamoid* bones. The patella is flattened and triangular in shape with the apex downwards. The posterior surface is smooth and articulates with the condyles of the femur; the anterior surface is roughened and is separated from the skin by a sac, similar to synovial membrane, called a *bursa*.

The *tibia* is the stronger of the two bones of the leg and lies on the inner, or medial, side. The upper end is greatly expanded to provide a good bearing surface of the body weight. It has two prominent masses, called the *medial* and *lateral* condyles which are smooth and articulate with the condyles of the femur. Between the condyles is a roughened area which gives attachment to the ligaments and cartilages of the knee joint. Below the condyles is a smaller projection called the *tuberosity of the tibia* which gives attachment to the *ligamentum patellae*. The lateral condyle has a small circular facet for articulation with the upper end of the fibula.

The *shaft* of the tibia is roughly triangular in cross-section. The *anterior border* lies immediately under the skin and can be felt as the shin. A second border faces the fibula and gives attachment to the *interosseous membrane* which connects tibia and fibula, as the ulna and radius are connected in the forearm.

The *lower end* of the bone is slightly expanded and projects downwards to form the *medial malleolus*, on the inner side of the ankle,

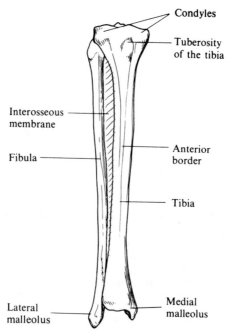

FIG. 63. The tibia and fibula (front view).

which articulates with the talus. The bone also articulates with the fibula.

The *fibula* is very slender compared with the tibia and lies on the outer side of the leg. The *head* of the fibula has a circular facet which articulates with the lateral condyle of the tibia but does not enter into the formation of the knee joint. The *shaft* is slender and ridged and one of the ridges gives attachment to the interosseous membrane connecting the tibia and fibula. The *lower end* of the fibula projects downwards to a lower level than the tibia and carries the bony prominence on the outside of the ankle known as the *lateral malleolus* which articulates with the talus.

The *tarsal bones* (or tarsus) comprise seven bones which make up the posterior half of the foot. The *talus* is the principal connecting link between the foot and the leg and forms an important

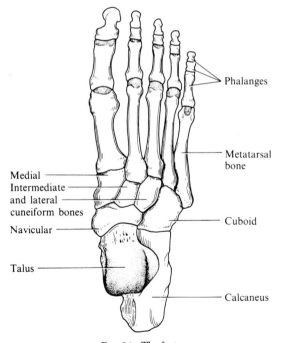

Phalanges

Metatarsal bone

Medial
Intermediate
and lateral
cuneiform bones

Navicular

Cuboid

Talus

Calcaneus

Fig. 64. The foot.

part of the ankle joint. The *calcaneus* is the largest and strongest of the tarsal bones; it projects backwards to form the prominence of the heel and to provide a lever for the muscles of the calf which are inserted into its posterior surface. The *navicular bone* lies between the talus and the three cuneiform bones. The three *cuneiform bones* are wedge-shaped and articulate with the navicular and the first three metatarsal bones. The *cuboid bone* is situated between the calcaneus and the fourth and fifth metatarsal bones.

The *metatarsal bones* (or metatarsus) are five miniature long bones, resembling the metacarpus. The *bases* articulate with the cuneiform and the cuboid bones. The *heads* articulate with the phalanges. As in the hand the *phalanges* of the foot correspond in number and arrangement, there being only two in the big toe and three in each of the other toes.

The Arches of the Foot

The foot has two main functions: to support the weight of the body and to propel the body forward when walking. To fulfil these functions the foot has two longitudinal arches, the *medial arch* which is particularly resilient and the *lateral arch* which is strong and allows only limited movement. There is also a series of *transverse arches*.

The arches of the foot give spring to the walk. They are helped in this by the strong ligaments, tendons and muscles which cross the sole of the foot and which, if stretched, may allow lowering of the medial longitudinal arch and eventually changes in the bones themselves, a condition known as 'flat foot'.

9 Joints or Articulations

A joint or articulation is formed wherever two bones meet, but not all junctions of bones allow movement. There are three groups of joints:

1. *Fibrous* or fixed joints
2. *Cartilaginous* or slightly movable joints
3. *Synovial* or freely movable joints

However not all joints fit rigidly into this classification as there are some fibrous joints which are slightly movable (e.g. the joint between the lower ends of the tibia and fibula) and some cartilaginous joints which are barely movable (e.g. the symphysis pubis).

Bones are joined to each other by ligaments, which are usually strong cords of fibrous tissue attached to the periosteum and running from one bone to another. These ligaments are yielding but inelastic and vary in strength and shape according to the work that they have to do. Because they are pliable they allow movement to take place but they are also strong, inelastic and rich in sensory nerves and in this way they protect the joints from excessive movement and strain.

Fibrous Joints

One type of fibrous joint occurs where the margins of two bones meet and dovetail accurately into one another, separated only by a thin band of fibrous tissue. Joints such as these, which do not normally permit movement, are found between the bones of the cranium and are called *sutures*. In the infant at birth there is a

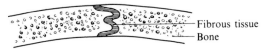

FIG. 65. Diagram of a fibrous joint.

definite line of fibrous tissue between two adjoining bones which allows the edges to glide over one another, enabling the head to be moulded to ease its passage through the birth canal. Other fibrous joints occur where the roots of the teeth articulate with the upper and lower jaws and where there is an interosseous ligament, as in the tibiofibular joint.

Cartilaginous Joints

A cartilaginous joint occurs where the two bony surfaces are covered with hyaline cartilage and connected by a pad of fibrocartilage and by ligaments, which do not form a complete capsule round the joint. A limited degree of movement is possible because the cartilaginous pad can be compressed. The joints between the bodies of the vertebrae and between the manubrium and the body of the sternum are cartilaginous joints.

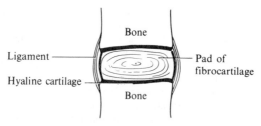

Fig. 66. Diagrammatic section through a cartilaginous (slightly movable) joint.

Synovial Joints

A synovial joint consists of two or more bones the ends of which are covered with articular hyaline cartilage. There is a joint cavity containing *synovial fluid* which nourishes the avascular articular cartilage and the joint is completely surrounded by a *fibrous capsule* lined with *synovial membrane* which lines the whole of the interior of the joint with the exception of the bone ends, menisci and discs. The bones are also connected by a number of ligaments, and some movement is always possible in a synovial joint even though it may be limited, as in the gliding movement between the adjoining metacarpal bones.

In some synovial joints the cavity may be divided by an *articular*

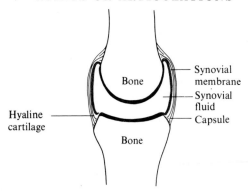

FIG. 67. Diagrammatic section through a synovial (freely movable) joint.

disc or *meniscus* consisting of fibrocartilage which helps to lubricate the joint, to reduce wear of the articular surfaces and to deepen the joint.

Types of Synovial Joints. The synovial joints are divided into several classes according to the type of movement which occurs.

1. *Hinge joints*, which allow movement in one direction only; the elbow is a good example.

2. *Pivot joints*, which only allow rotation, e.g. the radius and ulna at the elbow and the first and second cervical vertebrae allowing turning movements of wrist and head respectively.

3. *Condylar joints*, in which two pairs of articular surfaces allow movement in one direction only but the articular surfaces may be enclosed in the same or in different capsules; the knee joint is one example.

4. *Ball-and-socket joints* formed by a hemispherical head fitting into a cup-shaped socket. Examples are the hip and shoulder joints.

5. *Plane joints* in which gliding movements are restricted by ligaments or bony prominences, as in the carpal and tarsal joints.

Joint Movements

Joint movements are of three kinds: gliding, angular and circular. These movements are often combined to produce a variety of movement at any one joint.

Gliding movement occurs as its name implies without any angular or rotatory movement.

Angular movement brings about increase or decrease of the angles between the bones. It includes *flexion* or bending and *extension* or straightening, and also *abduction*, movement away from the midline and *adduction*, movement towards the midline.

Circular movement allows *internal rotation*, turning a part on its own axis towards the midline, and *external rotation* away from the midline. *Circumduction* allows movement of a limb through a circle. The words *supination* and *pronation* refer to turning the hand palm uppermost or palm downwards respectively.

Different types of joints have varying kinds of movement. The ball-and-socket joints are most freely movable allowing all movements described above, excluding supination and pronation. Hinge joints allow flexion and extension only. In gliding joints there is only slight movement increasing the range in all directions, as in the carpal and tarsal bones at wrist and ankle. In the hand the terms adduction and abduction are used to refer to movement to and from the middle line of the part and not from the midline of the body as a whole. Adduction of the thumb brings it to and across the palm of the hand, and adduction of the little finger brings it towards the thumb. In the anatomical position, with the palm facing forward, the little finger is moving away from the midline of the body but towards the midline of the hand.

The joints are movable, but the movements are carried out by the various muscles with which the joints are provided. The muscles, in addition to producing movement, run, as do the ligaments, from bone to bone, and help to hold the bones in position and give support to the joint capsule, as long as their normal tone is sustained. When the muscles are paralysed and limp, the looser joints are dislocated comparatively easily, particularly the freely movable shoulder joint; when the muscles are paralysed, but rigid and shrunken, the joint may become completely immovable. This immobility can be prevented by moving the joint through as wide a range of movement as possible to maintain the elasticity of the muscles. Immobility may also result from the joint surfaces becoming adherent to one another or to the joint capsule as a result of disease or injury affecting the muscle itself.

The Joints of the Head

The *temporomandibular joint*, between the temporal bone and the head of the mandible, is the only movable joint of the head and it is peculiar in that movement can occur in all three planes: upwards and downwards, backwards and forwards and from side to side.

The sutures between the skull bones have already been described. However, at the angles of the parietal bones there are unossified membranous areas called *fontanelles*.

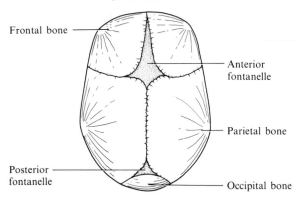

Frontal bone

Anterior fontanelle

Parietal bone

Posterior fontanelle

Occipital bone

FIG. 68. An infant's head at birth, showing the fontanelles.

The *anterior fontanelle* is the largest and lies at the junction of the two parietal bones with the frontal bone. It is diamond-shaped and does not close completely till the age of fifteen to eighteen months. Dehydration in infancy causes the fontanelle to become depressed which is a serious sign. A large venous sinus runs under the fontanelle, from which it is possible to obtain a specimen of blood and into which intravenous fluids may be given in the infant.

The *posterior fontanelle* lies at the junction of the parietal bones with the occipital bone. It is triangular in shape and closes shortly after birth. Delay in the closure of the fontanelles may be caused by hydrocephalus but also occurs sometimes in the normal child.

The Joints of the Trunk

There are joints between all the vertebrae from the second cervical to the sacrum. Cartilaginous joints lie between the vertebral bodies and synovial joints between the vertebral arches. Because the joints are so numerous the spinal column as a whole has considerable movement. *Anterior* and *posterior longitudinal ligaments* pass from the top of the spine to the sacrum giving support. Other ligaments pass between the vertebral arches.

Between the ribs and the vertebrae are the *costovertebral joints* which allow gliding movements; the same type of movement occurs at the *sternocostal joints*.

The Joints of the Upper Extremity

The *sternoclavicular joint* is formed by the sternal end of the clavicle, the manubrium of the sternum and the cartilage of the first rib. It allows a gliding movement of the clavicle.

The *acromioclavicular joint* lies between the acromial extremity of the clavicle and the acromion of the scapula and is usually associated with movements of the shoulder.

The *shoulder joint* is a ball-and-socket joint, and the most freely movable of the joints in the body. It is formed by the head of the humerus fitting into the small, shallow glenoid cavity. The articular surfaces are covered with articular cartilage and the glenoid cavity is enlarged and deepened by a rim of fibrocartilage, called the *glenoidal labrum*, which runs around it. This lessens the risk of dislocation without limiting the movement as much as a larger and deeper bony socket would do. The bones are held together by a loose capsule of ligaments to give the joint its wide range of movement, but the powerful muscles help to keep the bones in position. The long tendon of the biceps muscle serves as an intracapsular ligament. It runs

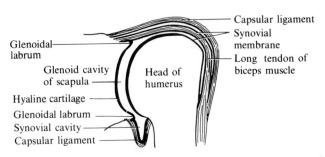

Fig. 69. Diagrammatic section through the shoulder joint.

through the bicipital groove between the tuberosities of the humerus into the joint cavity and, since it arises from the scapula immediately above the glenoid cavity, it tends to hold the articular surfaces in position. Movement of the arm above the level of the shoulder joint is due to movement of the scapula over the back of the thorax.

The *elbow joint* is complicated by the fact that it contains the hinge joint between the humerus and the radius and ulna and the pivot joint between the radius and ulna. There are strong ligaments running between the three bones and a circular ligament called the *annular ligament* which keeps the head of the radius in contact with the radial notch of the ulna. The lower end of the radius also forms a pivot joint with the ulna.

The *wrist joint* is formed by the lower end of the radius and the scaphoid, lunate and triquetral bones. Together with the joints between the carpal bones the movements of flexion, extension, adduction (ulna deviation), abduction (radial deviation) and circumduction can be carried out.

The *metacarpophalangeal joints* are also capable of all the movements described for the wrist joints but the interphalangeal joints are hinge joints, allowing flexion and extension only.

The Joints of the Lower Extremity

The *sacroiliac joint* is a synovial joint which allows a small amount of rotatory movement during flexion and extension of the trunk.

The *symphysis pubis* is a cartilaginous joint which moves very little. However during pregnancy the pelvic joints and ligaments are relaxed to allow slightly greater movement.

The *hip joint* is a ball-and-socket joint formed by the head of the femur fitting into the cup-shaped acetabulum. The joint surfaces are covered with articular cartilage and the acetabulum, like the glenoid cavity, is deepened by a rim of fibrocartilage called the *acetabular labrum*. The *ligament of the head of the femur* is attached to a small, roughened pit called the *fovea* near the centre of the head of the femur, and runs to the acetabulum. The joint has a strong *fibrous capsule* and many ligaments, one of which, the *ilio-femoral ligament*, lies across the front of the joint and prevents extension of the hip beyond a straight line with the trunk. A wide range of movement is possible at the hip joint, though when the knee is flexed, flexion of the hip is limited by the contact of the thigh with

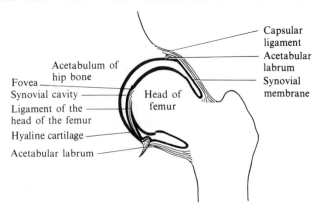

FIG. 70. Diagrammatic section through the hip joint.

the anterior abdominal wall; also, when the knee is extended, flexion
of the hip is limited by tension of the hamstring muscles.

The *knee joint* is the largest joint of the body. It is a compound
joint: a condylar joint which gives articulation between the condyles
of the femur and the tibia and a plane joint which gives articulation

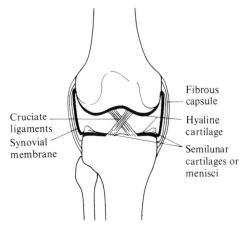

FIG. 71. Diagrammatic section through the knee joint.

between the patella and the femur. The joint has a *fibrous capsule* into the front of which the patella enters and which is lined with synovial membrane. The *cruciate ligaments* are very strong and cross each other within the joint. They run from the intercondylar area of the tibia to the femur and are partially covered with synovial membrane. The extracapsular ligaments are also strong and thick and help in controlling movement of the joint. The *menisci* (or semilunar cartilages) deepen the surfaces of the upper end of the tibia. They are wedge-shaped, the outer border being thick and convex and the inner thin and concave, and they can be injured as a result of a twisting strain when the knee is flexed. However, if fully exercised they will reform. The movements of the knee joint are mainly flexion and extension, although some rotation can also occur.

The *upper tibiofibular joint* is a synovial plane joint allowing a little gliding movement while at the lower end of the bones there is a small amount of rotation of the fibula during some ankle movements.

The *ankle joint* is a hinge joint formed between the tibia, fibula and talus. The movements are flexion and extension but are usually

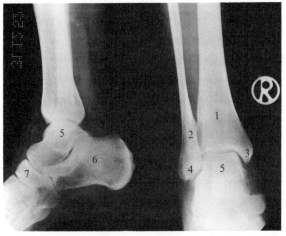

Fig. 72. X-rays of the right ankle joint, showing (left) lateral view, and (right) antero-posterior view. (1) The tibia (2) the fibula, (3) the medial malleolus, (4) the lateral malleolus, (5) the talus, (6) the calcaneum, and (7) the cuboid.

spoken of as *dorsiflexion*, raising the foot, and *plantar flexion*, raising the heel.

The joints between the tarsal bones and between the tarsus and metatarsus are gliding joints and movement is limited. The *metatarsophalangeal joints* and the *interphalangeal joints* allow movements similar to the corresponding joints in the hand.

10 Structure and Action of Muscle

Structure of Muscle

The muscular system consists of a large number of muscles through which the movements of the body are carried out. Voluntary muscles are attached to bones, cartilages, ligaments, skin or to other muscles by fibrous structures called *tendons* and *aponeuroses*. The individual fibres of voluntary muscle with their sheaths of *sarcolemma* are bound together into bundles by the *endomysium* and are covered by the *perimysium*. The bundles, or *fasciculae*, are bound together by a denser covering called the *epimysium* and these groups form the individual voluntary muscles of the body (see Fig. 22). All muscles have a good blood supply from nearby arteries. Arterioles in the perimysium give off capillaries which run in the endomysium and across the fibres. Blood vessels and nerves enter the muscle together at the hilum.

Most muscles have tendons at one or both ends. *Tendons* are made of fibrous tissue and are usually cord-like in appearance, though in some flat or sheet-like muscles the cord is replaced by a thin, strong fibrous sheet called an *aponeurosis*. Fibrous tissue also forms a protective covering or muscle sheath, known as fascia.

Where one muscle is attached to another the fibres may interlace, the perimysium of one fusing with the perimysium of the other, or the two muscles may share a common tendon. A third type of connection occurs in the muscles of the abdominal wall where the fibres of the aponeuroses interlace, forming the *linea alba*, which can be seen as a shallow groove above the umbilicus.

Action of Muscle

When a muscle contracts one end normally remains stationary while the other end is drawn towards it. The end which remains stationary is called the *origin* and that which moves is called the *insertion*. It is

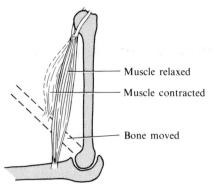

FIG. 73. Diagram showing bone movement as muscle contracts.

not uncommon, however, for a muscle to be used, as it were, the wrong way round so that the insertion remains fixed and the origin moves towards it. The gluteus maximus provides an illustration. Its origin is in the sacrum and it is inserted into the femur. When the insertion moves towards the origin the flexed thigh is extended; when the body is bent forward at the hips the standing position is regained by movement of the origin towards the insertion. This arrangement economizes on the number of muscles required and further economy is achieved by the placing of muscles so that they can carry out more than one action. Muscles must cross the joint they move; some cross two joints producing movement in both, e.g. the biceps crosses both elbow and shoulder causing flexion of both.

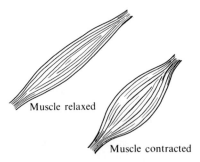

FIG. 74. A typical voluntary muscle showing contraction.

Muscles only act by contracting and pulling; they cannot push, though they can contract without shortening and so hold a joint firm and fixed in a certain position. When the contraction passes off the muscles become soft but do not lengthen until stretched by the contraction of opposing muscles, known as antagonists.

Muscles never work alone; even the simplest movement requires the action of many muscles. Picking up a pencil requires movement of fingers and thumb, wrist and elbow and possibly of shoulder and trunk as the body leans forward. Each muscle must contract just sufficiently and each antagonist relax equally to allow the movement to take place smoothly without jerking. This concerted action of many muscles is termed *muscle coordination*; any new action involving coordination requires time and practice until the new combination of muscle movement has been acquired and only then can it be carried out without great mental effort and concentration.

The sensory nerve gives 'muscle sense', not a very acute sensation but sufficient to allow awareness of contraction and relaxation in muscle. There is no awareness of this sensation until a conscious effort is made to relax or contract a muscle at which time the previous degree of concentration becomes obvious. Under normal circumstances the muscles are in a state of partial contraction known as *muscle tone*; it is because of muscle tone that a position can be maintained for long periods without exhaustion. This is dependent on a mechanism whereby different groups of muscle fibres contract and relax in turn giving periods of rest and activity to each group. The muscles having the highest degree of tonicity in man are those of the neck and back.

Contraction of Muscle

Muscle is composed of:

75 per cent water
20 per cent protein
5 per cent mineral salts, glycogen and fat

Muscle contraction occurs as a result of nerve impulses. In order for muscle fibres to contract energy is required and this is obtained from the oxidation of food, particularly carbohydrates. During digestion carbohydrates are broken down to a simple sugar called glucose. The glucose which is not required immediately by the body is converted to glycogen and is stored in the liver and in the muscles. Muscle glycogen provides the source of heat and energy for muscular activity. During the oxidation of glycogen to carbon dioxide and water a compound is formed which is rich in energy. This compound is called adenotriphosphate (ATP). When it is necessary for muscular contraction the energy from ATP can be released as the compound changes to adenodiphosphate (ADP). During the oxidation of glycogen pyruvic acid is formed. If plenty of oxygen is available, as it usually is during ordinary movement, pyruvic acid is broken down to carbon dioxide and water, and during the process energy is released, which is used to make more ATP. If insufficient oxygen is available the pyruvic acid is converted to lactic acid, which accumulates and produces muscle fatigue.

During violent exercise more oxygen is brought to the muscles but even so not enough oxygen reaches the muscle cells, particularly at the beginning of the effort. Lactic acid accumulates and diffuses into the tissue fluid and the blood. The presence of lactic acid in the blood stimulates the respiratory centre and the rate and depth of respiration are increased. This continues, even after the exercise is over, until sufficient oxygen has been taken in to allow the cells of the muscles and the liver to oxidize the lactic acid completely, or convert it to glycogen. This extra oxygen needed to remove the accumulated lactic acid is called the 'oxygen debt' which must be repaid after the exercise is completed.

Table 2

Changes during Muscle Contraction

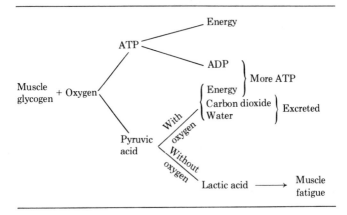

11 The Chief Muscles of the Body

The Muscles of the Head

The muscles of the head are divided into two groups according to their functions: (1) the muscles of expression, and (2) the muscles of mastication.

The *muscles of expression* are attached to the skin rather than to the bone, so that they move the skin and change the facial appearance. Circular muscles, called *orbicularis oculi* and *orbicularis oris*, surround the eyes and mouth respectively, closing them.

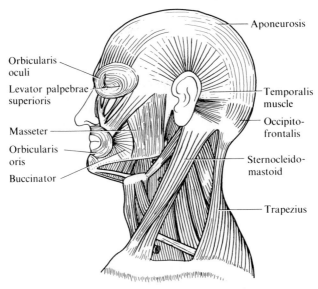

FIG. 75. The muscles of the head and neck.

Small muscles raise and lower the eyebrows and upper lids, raise and lower the angles of the mouth and dilate the nostrils, causing a look of surprise, worry, happiness or sorrow. Small muscles also move the eyeballs in the orbits (see Chapter 24) both to direct the eyes for sight and also to change the expression.

The *muscles of mastication* move the lower jaw up and down in biting, and also both from side to side and backwards and forwards in chewing. They are the *masseter* running to the angle of the jaw from the zygomatic arch, the *temporalis* muscle, lying over the temporal bone and inserted into the lower jaw, and the smaller muscles which also run from the skull to the lower jaw. These are the muscles which 'lock the jaw' in tetanus.

The Muscles of the Neck

The neck contains two large muscles, the sternocleidomastoid and the trapezius.

The *sternocleidomastoid* lies at the side of the neck, running from the sternum and clavicle to the mastoid process and the surface of the temporal bone behind it. When the muscle on one side contracts it draws the head towards the shoulder and turns the head. When both are used together they flex the neck.

The *trapezius* lies over the back of the neck and shoulder and is roughly triangular in shape, with the base joining the spine down the back of the neck and chest, from the occiput, to which it is also attached, downwards; the angle is inserted into the scapula and clavicle, over the top and back of the shoulder. It draws the shoulders back when used as a whole, and also draws the scapula up and down, when the upper and lower portions are used separately (Table 3).

The Muscles of the Trunk

The chief muscles of the trunk can be grouped according to their function into:

1. Muscles moving the shoulder
2. Muscles of respiration
3. Muscles forming the abdominal wall
4. Muscles moving the hip
5. Muscles moving the spine
6. Muscles of the pelvic floor

Table 3
Muscles of the Neck

Name	Position	Origin	Insertion	Action
Sternocleidomastoid	The side of the neck	Sternum and clavicle	Mastoid process	Used separately, turn the head to the side and tilt to same shoulder; used together, flex the neck
Trapezius	The back of the neck and shoulder	The occiput and the spines of the thoracic vertebrae	The clavicle and the spine of the scapula	Draws the scapula back, bracing the shoulders; raising and lowering the shoulder and extending the neck by exerting pull on the occiput

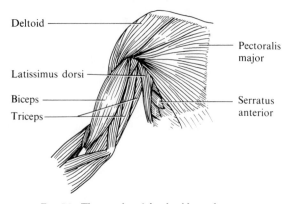

FIG. 76. The muscles of the shoulder and arm.

Muscles Moving the Shoulder

The chief muscles moving the shoulder are the powerful muscles covering the back and front of the chest. They include the pectoralis major, the trapezius (see p. 100), the latissimus dorsi and the serratus anterior. The *pectoralis* covers the front of the chest, running from the sternum out to the humerus. The *latissimus dorsi* covers the back of the chest and abdomen, running from the thoracic and lumbar vertebrae and iliac crest out to the humerus. These muscles form the muscle in front of and behind the armpit. The *serratus anterior* runs round the side wall of the thorax from the ribs in front to the vertebral edges of the scapula under which it passes (Table 4).

Muscles of Respiration

The chief muscles of respiration are:

1. The diaphragm
2. The external intercostals
3. The internal intercostals

The *diaphragm* is a dome-shaped sheet of muscle dividing the thorax from the abdomen. The border is of muscle, while the centre is a sheet of fibrous tissue or aponeurosis. The muscle arises from the tip of the sternum, the lower ribs and their cartilages, and the first three lumbar vertebrae, and is inserted into the central aponeurosis.

Table 4

Muscles Connecting the Upper Limb with the Trunk

Name	Position	Origin	Insertion	Action
Pectoralis major	Front of chest	Sternum, clavicle and cartilages of true ribs	Humerus (bicipital groove)	Adducts the shoulder, draws the arm across the front of the chest. Internal rotation of shoulder
Latissimus dorsi	Crosses the back from the lumbar region to the shoulder	Lower thoracic, lumbar and sacral vertebrae and iliac crest	Humerus (bicipital groove)	Adducts the shoulder, draws the arm backwards and downwards, as in bell-pulling and rowing, and internal rotation of shoulder
Serratus anterior	Over the sides of the thorax and under the scapula at the back	The front of the eight upper ribs	The medial border of the scapula	Draws the scapula forwards; antagonistic to the trapezius

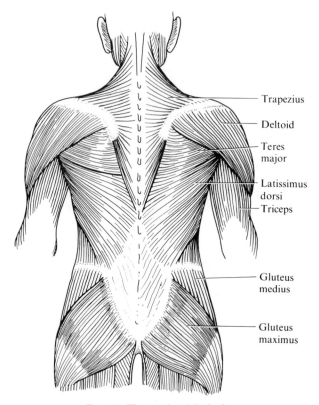

FIG. 77. The muscles of the back.

Three openings in the diaphragm allow for the passage of the oeso-phagus, aorta and inferior vena cava, together with lesser structures such as the vagus nerve and thoracic duct, which accompany the oesophagus and aorta respectively. When the muscle fibres contract the dome of the diaphragm is flattened and lowered, which increases the depth of the thorax from top to bottom.

The *external intercostal* muscles lie between the ribs, their fibres running downwards and forwards from one rib to the rib below. Each originates from the lower border of the rib above and is

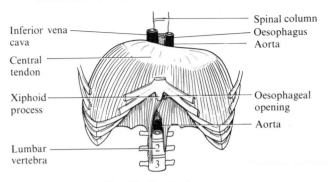

Inferior vena cava

Central tendon

Xiphoid process

Lumbar vertebra

Spinal column

Oesophagus

Aorta

Oesophageal opening

Aorta

FIG. 78. The diaphragm.

inserted into the upper border of the rib below. Their action is to draw the ribs upwards and outwards to increase the size of the thorax from side to side and from back to front.

The *internal intercostal* muscles also lie between the ribs under the external intercostals and are antagonistic to them. Their fibres run downwards and backwards from one rib to the rib below. Each originates from the lower border of the rib above and is inserted into the upper border of the rib below. Their action is to draw the ribs downwards and inwards to decrease the size of the thorax from side to side and from back to front, particularly in forced expiration.

Muscles Forming the Abdominal Wall

The chief muscles of the abdominal wall are:

1. The rectus abdominis, forming the front wall
2. The external oblique
3. The internal oblique } forming the side wall and
4. The transversus abdominis } lying one under the other
5. The quadratus lumborum

The *rectus abdominis* forms the anterior abdominal wall, running up from the pubis to the sternum and costal cartilages. Its fibres run straight up and down, hence its name, for rectus means straight. It is divided into two parts by a line of fibrous tissue in the midline of the body called the *linea alba*. It is also crossed by lines of fibrous tissue at intervals. These fibrous bands strengthen the muscle and help to prevent stretching.

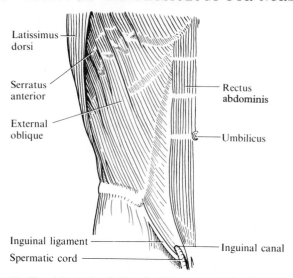

Fig. 79. The abdominal wall. Note that the aponeurosis has been cut away to expose the rectus.

The *external oblique* muscle forms the outer layer of the side wall. Its fibres run downwards and forwards. It arises from the lower ribs and is inserted into the iliac crest and the inguinal ligament. The *inguinal ligament* forms the firm edge of the abdominal wall across the groin, where the muscles are not inserted into the bone, leaving a gap for muscles, blood vessels and nerves to pass under it into the limb from the trunk. It gives attachment to the muscles here. It is a strong cord of fibrous tissue. Across the front of the abdomen the *external oblique* forms a strong aponeurosis which passes in front of the rectus, joining the linea alba.

The *internal oblique* forms the second layer of the side wall of the abdomen. Its fibres run upwards and forwards. It arises from the iliac crest and the inguinal ligament and is inserted into the lower ribs and their cartilages. It also forms an aponeurosis, passing partly in front and partly behind the rectus, and joining those of the external oblique and transversus abdominis.

The *transversus abdominis* forms the inner layer of the side wall of the abdomen, lying under the internal oblique. Its fibres run straight round the abdominal wall. It arises from the iliac crest and the lumbar fascia, by which it is joined to the lumbar vertebrae. It is inserted into an aponeurosis which runs across the front of the abdomen, behind the rectus and joins the linea alba.

The *quadratus lumborum* forms the posterior wall, running up from the iliac crest to the twelfth rib and upper lumbar vertebrae. It holds the twelfth rib steady during breathing.

The abdominal wall is pierced by a canal in either groin; this is called the *inguinal canal*. It runs obliquely through the muscular coats just above the inguinal ligament near its inner extremity. The canal gives passage to structures: in the male the spermatic cord from the testicle, in the female the round ligaments of the uterus run through it with associated blood vessels and nerves.

Muscles of the Hip

The muscles in the trunk moving the hip are:

1. The iliopsoas
   a. The psoas
   b. The iliacus
2. The gluteal muscles: maximus, medius, minimus

The *iliopsoas* muscles cross the front of the groin behind the inguinal ligament. The *psoas* arises from the transverse processes of the lumbar vertebrae, and the *iliacus* from the front surface of the upper part of the ilium. They are both inserted into the lesser trochanter of the femur. They flex the hip joint and also produce abduction but when the femur is fixed, they bend the trunk forwards.

The *gluteal muscles* form the buttocks (see Fig. 77), running from the back of the sacrum and ilium to their insertion in the greater trochanter of the femur and the gluteal ridge below it. They are three in number: the gluteus maximus, medius and minimus. They extend the hip joint and also abduct and laterally rotate the hip (Table 5) but when the femur is fixed, they extend the trunk on the lower limb. These muscles are commonly used as the site for intramuscular injections as they are thick and fleshy. Care must be taken to use the upper outer quadrant as the sciatic nerve passes through the other quadrants.

Muscles Moving the Spine

The muscles of the abdominal wall flex and turn the trunk, the

Table 5

Trunk Muscles which Move the Hip

Name	Position	Origin	Insertion	Action
Psoas major	Crosses the groin behind the inguinal ligament	Transverse processes of lumbar vertebrae	Lesser trochanter of femur	Flexes the hip
Iliacus	Crosses the groin behind the inguinal ligament with the psoas	Front surface of the iliac bone	Lesser trochanter of femur	Flexes the hip
Gluteal muscles	Cross the back of the hip, forming the buttocks	Posterior surface of ilium and sacrum	Greater trochanter and gluteal line of femur	Extend, abduct and laterally rotate the hip

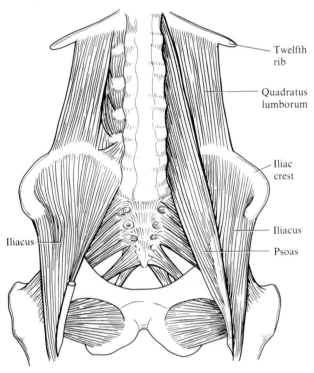

Twelfth
rib

Quadratus
lumborum

Iliac
crest

Iliacus

Psoas

Iliacus

FIG. 80. The iliopsoas.

rectus flexing and the side muscles turning the thorax on the
abdomen. The abdominal muscles also compress the internal organs.
The *spinalis muscles* extend the spine. They run up the back of the
trunk on either side of the spine. They arise from the back of the
iliac crest and the sacrum, and are inserted into the ribs and upper
vertebrae.

Muscles of the Pelvic Diaphragm

The pelvic diaphragm consists of muscles which form the support
of the pelvic organs; it runs from the pubis in front back to the
sacrum and coccyx and out to the ischium on either side. It is like an

open book in shape, sloping downwards from back to front and from either side towards the midline. It is composed of the *levator ani* and *coccygeus* muscles. It is pierced in the midline in the female by three openings for the passage of the urethra, the vagina and the rectum. In the male there only two openings, for the urethra and the rectum respectively.

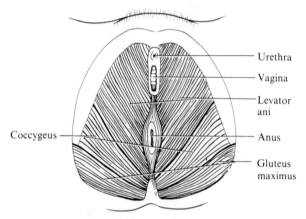

FIG. 81. The pelvic diaphragm.

The Muscles of the Upper Limb

The muscles of the upper limb may be divided into the muscles of the arm, the forearm and the hand.

The Muscles of the Arm

The muscles of the arm are the largest and strongest of the limb and include:

1. The biceps
2. The triceps
3. The deltoid
4. The brachialis

The *biceps* is so called because it has two heads (Latin caput — head). It runs down the front of the arm, where it can easily be felt when it is contracted. It arises by two heads, one from the glenoid cavity and one from the coracoid process of the scapula, and is inserted into the radial tuberosity in the forearm, crossing the front of the elbow joint. It flexes the elbow and also the shoulder and supinates the hand. Hence the act of supination can be carried out with great force and screws and nuts are made so that a right-handed man tightens them with the action of supination.

The *triceps* is so called because it has three heads. It lies at the back of the arm; it arises by three heads, one from the scapula and two from the humerus, and is inserted into the olecranon of the ulna, at the back of the elbow joint. It extends both elbow and shoulder and is antagonistic to the biceps.

M   A   L

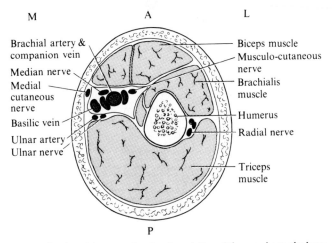

Brachial artery &
companion vein

Median nerve

Medial
cutaneous
nerve

Basilic vein

Ulnar artery
Ulnar nerve

Biceps muscle

Musculo-cutaneous
nerve

Brachialis
muscle

Humerus

Radial nerve

Triceps
muscle

P

FIG. 82. Section of the arm, showing the relation of the muscles to the bone and other structures. L, lateral; M, median; A, anterior.

The *deltoid* is triangular in shape. It lies over the shoulder in the position of an epaulette with the base of the triangle forming the origin and attached to the shoulder girdle, just above the shoulder joint. It is inserted into the deltoid tuberosity on the outer side of the humerus. Its action is to abduct the shoulder to a right angle. (To raise the arm above a right angle, the shoulder girdle must move too, and the trapezius comes into play, drawing the scapula and clavicle up towards the occiput.) The front part of the deltoid muscle, used alone, helps to flex the shoulder, moving the humerus forwards, and the back portion, used alone, helps to extend the shoulder, moving the humerus backwards and towards the midline.

The *brachialis* lies lower in the front of the arm than the biceps, arising from the humerus and being inserted into the coronoid process of the ulna. It assists the biceps in the powerful action of flexion of the elbow (Table 6).

The Muscles of the Forearm

The forearm contains numerous small, less powerful muscles for movement of the wrist and digits. In the front lie the flexors of the

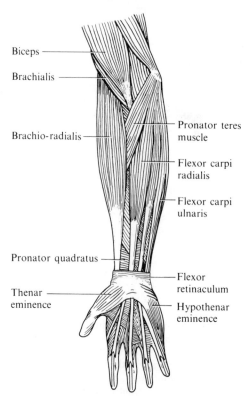

Biceps

Brachialis

Brachio-radialis

Pronator quadratus

Thenar eminence

Pronator teres muscle

Flexor carpi radialis

Flexor carpi ulnaris

Flexor retinaculum

Hypothenar eminence

Fig. 83. The muscles of the forearm and hand.

Table 6
Muscles of the Arm

Name	Position	Origin	Insertion	Action
Biceps (two heads)	The front of the arm	Coracoid process and above the glenoid cavity of the scapula	Radial tuberosity	Flexes elbow and shoulder and supinates the hand
Triceps (three heads)	The back of the arm	One head from the axillary border of the scapula; two from the shaft of the humerus	Olecranon of ulna	Extends elbow and shoulder
Deltoid	Over the shoulder	Acromion and spine of scapula; the clavicle	Deltoid tuberosity of humerus	Abducts the shoulder to a right angle
Brachialis	Crosses the front of the elbow	Humerus	Anterior surface of coronoid process of ulna	Flexes the elbow

wrist, the common flexors of the fingers, the long flexor of the thumb and the pronator muscles of the wrist. The common flexors of the digits divide into four tendons, which run across the palm of the hand and up to the terminal phalanx of each digit, into which they are inserted. At the back lie the extensors of the wrist, the common extensors of the fingers, the extensor of the thumb and of the first finger, and the supinators of the wrist. The tendons of the muscles which cross the wrist are bound down by the *flexor retinaculum* just above the wrist joint. In the same way the tendons are bound down to the fingers to keep them close to the bones.

The Muscles of the Hand

The hand contains very little muscle, as this would tend to make it clumsy and interfere with its usefulness in grasping and lifting. Many of the muscles which move it are therefore in the forearm. The hand only contains the short flexor of the thumb, and the adductor and abductor muscles for the digits. These latter are termed the interosseous muscles, and are only well developed at the base of the thumb and to a lesser extent at the base of the little finger. Here they form the *thenar* and *hypothenar eminences*, and give power to the grip, where adduction of the thumb is particularly important.

The Muscles of the Lower Limb

The muscles of the lower limb are much larger and more powerful than those of the upper, as the limb carries the whole weight of the body. They may be divided into the muscles of the thigh, the leg and the foot.

The Muscles of the Thigh

The muscles of the thigh are particularly strong and include:

1. The quadriceps femoris
2. The hamstrings
3. The sartorius
4. The adductors of the hip

The *quadriceps* is so called because it has four heads; rather it is four muscles with a common insertion into the patella, and through the patellar ligament it joins the tibia and is the extensor of the knee joint, used in standing and in the powerful action of kicking. It is made up of the *rectus* or straight muscle, and the three *vastus*

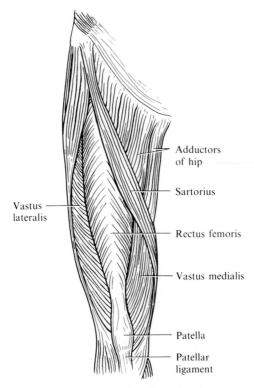

FIG. 84. The muscles of the front of the thigh.

muscles, the lateral, intermedial and medial. Of these the lateral is
the longest muscle and lies on the outside of the thigh. It is occasionally used for the giving of intramuscular injections, as it is well away
from the blood vessels, nerves and lymphatics of the limb here.

The *hamstrings* are the flexor muscles of the knee and are so
called because of the strong tendons or 'strings' that they form on
either side of the popliteal space, at the back of the knee joint. These
can be seen and felt readily when the knee is bent. The hamstring
muscles are:

1. The *biceps femoris*, so called because it arises by two heads,

one from the ischial tuberosity, the other from the back of the femur. It is inserted into the fibula. It lies on the outside of the back of the thigh.

2. The *semitendinosus* is so called bceause of the length of the tendon by which it is inserted into the tibia. It arises from the ischial tuberosity, with the biceps muscle, and lies in the middle of the back of the thigh.

3. The *semimembranosus* is so called because of the membranous tendon by which it arises from the ischial tuberosity. It is inserted into the tibia and lies on the inner side of the back of the thigh.

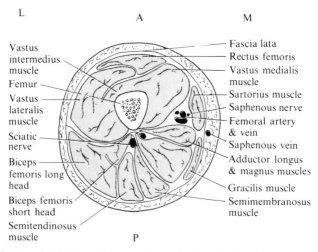

FIG. 85. Section through the thigh, showing the relation of the muscles to the bone and other structures. L, lateral; M, median; A, anterior; P, posterior aspect.

The biceps thus forms the 'hamstring' tendon on the outer side of the back of the knee and the semitendinosus and the semimembranosus form the 'hamstring' tendon on the inner side of the back of the knee. The hamstring muscles are again a very powerful group of muscles serving to bend the knee in walking, jumping and climbing, and also, when the tibia is fixed, as in standing, pulling on the ischial tuberosity and helping the gluteal muscles to extend the hip joint.

The *sartorius*, or tailor's muscle, runs from the superior anterior iliac spine across the front of the thigh to the inner side of the knee, which it crosses, and is inserted into the tibia. It assists in the joint movements which occur when sitting tailorwise, flexing hip and knee and rotating the femur.

The *adductor muscles* form the flesh on the inside of the thigh and are muscles by which the hip is adducted, being assisted by the smaller, more superficial muscle, the *gracilis*. These muscles are particularly well developed in persons who ride horses, since the rider holds on by the knees, keeping the hips adducted. The adductor

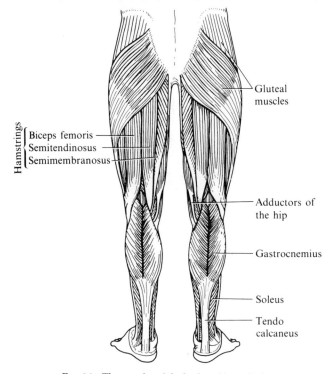

FIG. 86. The muscles of the back and lower limbs.

Table 7

Muscles of the Thigh

Name	Position	Origin	Insertion	Action
Quadriceps femoris	Front of thigh	Ilium and femur	Patella, through which it is joined to the tibia by the patellar ligament	Extension of knee and flexion of hip
Hamstrings	Back of thigh	Ischial tuberosity and femur	Tibia and fibula by tendons at either side of the popliteal space	Flexion of the knee and extension of the hip
Sartorius	Crosses front of thigh	Anterior, superior iliac spine	Tibia at inner side below the knee	Assists in flexion of hip and knee and abduction and external rotation of hip
Adductors of the hip	Inner side of thigh	Pubis and ischium	Linea aspera and medial condyle of femur	Adduct the hip

muscles run from the pubis and ischium and are inserted into the linea aspera and medial condyle of the femur. The adductor magnus forms the greater part of this muscle and a canal runs through it which allows the main artery of the thigh to run from the inner side of the thigh to the back of the knee, where it is well protected (Table 7).

The Muscles of the Leg

In the leg there are a few large muscles controlling the ankle and many smaller ones moving the foot. The chief muscles are:

1. The gastrocnemius ⎫
2. The soleus   ⎬ the calf muscles
3. The tibialis anterior
4. The flexors and extensors of the digits

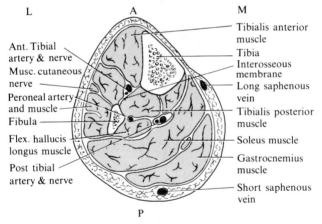

FIG. 87. Section through the leg, showing the position of the muscles in relation to the bones. A, the axis of the centre of the limb; M, median; L, lateral aspect.

The *gastrocnemius* and *soleus* together form the flesh of the calf, the gastrocnemius lying at the back and the soleus lying in front of it. The *gastrocnemius* arises by two heads from the femur, forming the borders of the popliteal space below, as the hamstrings do above the knee. The *soleus* arises from the tibia, not crossing the knee joint and thus not affecting its movements. Both muscles unite below to form a strong common tendon, the tendo calcaneus, by which they are inserted into the calcaneum. The calf muscles raise the heel, causing plantar flexion, or, as it is sometimes called today, extension, of the ankle joint, as in walking and running.

The *tibialis* lies in front of the leg just outside the crest of the tibia, where it can be seen and felt when the toes and ball of the foot are raised from the ground. It is this muscle which gets stiff from abnormal exercise, when persons unused to walking up steep slopes begin to climb mountains. It arises from the tibia and fibula below the knee joint, and is inserted into the tarsal and metatarsal bones

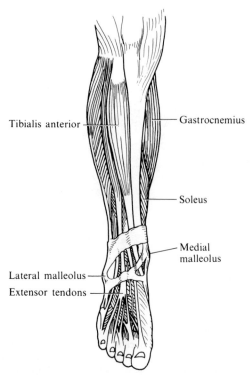

FIG. 88. The muscles of the front of the leg and foot.

Table 8
Muscles of the Leg

Name	Position	Origin	Insertion	Action
Gastrocnemius	The calf	Condyles of the femur	Calcaneum	Plantar flexion of the ankle; raising the heel
Soleus	The calf	Tibia and fibula	Calcaneum	
Tibialis anterior	Side of the leg	Tibia	Tarsals and meta-tarsals inside the instep	Dorsiflexion of the ankle, raising the toes; inversion of the foot

on the inside of the instep, where its tendon can be seen and felt readily when the ball of the foot is raised from the ground and the sole turned to face in towards the midline. It produces dorsiflexion of the ankle, i.e. bending towards the dorsal surface, or, as it is some-times called today, flexion, bringing the terminology into line with that used for all other joints. It also inverts the foot (Table 8).

The Muscles of the Foot

The foot, like the hand, contains little muscle, the chief muscles moving the foot lying largely in the leg. The tendons of the extensors of the digits cross the dorsal surface of the foot, the big toe having an individual muscle and tendon. The tendons of the flexors of the digits cross the sole and are strong and very important in helping to support the arch of the foot. There is a common flexor for the small toes and a flexor for the big toe. In addition, a short flexor of the toes crosses the sole from the calcaneum to the phalanges, and also gives support to the arch. Small interosseous muscles between the metatarsal bones abduct and adduct the digits, but are little used, and therefore little developed.

12 The Blood

The circulatory system is the transport system of the body by which food, oxygen, water and all other essentials are carried to the tissue cells and their waste products are carried away. It consists of three parts:

1. The *blood*, which is the fluid in which materials are carried to and from the tissues.
2. The *heart*, which is the driving force which propels the blood.
3. The *blood vessels*, the routes by which the blood travels to and through the tissues, and back to the heart.

The blood is a thick red fluid; it is bright red in the arteries, where it is oxygenated, and a dark purplish-red in the veins, where it is deoxygenated, having given up some of its oxygen to the tissues—the cause of the colour change—and received waste products taken in from them. It is slightly alkaline in reaction, and the reaction varies very little during life, as the cells of the body can live only if the reaction is normal.

It forms about five per cent of the body weight, so that the average volume is three to four litres.

Composition of the Blood

Although apparently fluid, blood actually consists of a fluid and a solid part. When examined under the microscope, large numbers of small round bodies can be seen in it which are known as the *blood corpuscles* or cells. These form the solid part, while the liquid in which they float is the fluid part and is called plasma. The cells form 45 per cent and the plasma 55 per cent of the total volume.

Plasma

The plasma or fluid part of the blood is a clear, straw-coloured, watery fluid, similar to the fluid found in an ordinary blister. It consists of:

1. Water, which forms over 90 per cent of the whole.

2. Mineral salts. These include chlorides, phosphates and carbonates of sodium, potassium and calcium. The chief salt present is sodium chloride. The correct balance of the various salts is necessary for the normal functioning of the body tissues, and there is a total of $0 \cdot 9$ per cent inorganic substances.

3. Plasma proteins: albumin, globulin, fibrinogen, prothrombin and heparin.

4. Foodstuffs in their simplest forms: glucose, amino acids, fatty acids and glycerol, and vitamins.

5. Gases in solution: oxygen, carbon dioxide and nitrogen.

6. Waste products from the tissues: urea, uric acid and creatinine.

7. Antibodies and antitoxins which protect the body against bacterial infection.

8. Hormones from the ductless glands.

9. Enzymes.

The *water* in plasma provides fresh water to supply the fluid which bathes all the body cells and renews the water within the cells. Sixty per cent of the body weight is water and in an average man weighing 70 kg (kilograms) this would be approximately 46 litres. Of the 46 litres approximately 29 litres are within the cells (intracellular fluid) and 17 litres are outside the cells (extracellular fluid). The extracellular fluid is divided between the blood vessels (3 litres) and the fluid bathing the cells, called the interstitial fluid (14 litres).

The *salts* in the plasma are necessary for the building of protoplasm and they act as buffer substances neutralizing acids or alkalies in the body and maintaining the correct reaction of the blood. Blood is always slightly alkaline in health and has a pH of $7 \cdot 4$ (see Chapter 1). In plasma there are approximately 155 mmol/litre of positively charged ions, chiefly sodium, balanced by 155 mmol/litre of negatively charged ions, mainly chloride and bicarbonate. This is referred to as the *electrolyte balance*, and is similar in the interstitial fluid. In the intracellular fluid potassium replaces sodium as the positively charged ions, and phosphate ions and proteins replace chloride as the negatively charged ions.

The *proteins* which plasma contains give the blood the sticky consistency, called viscosity, which is necessary to prevent too much fluid passing through the capillary walls into the tissues. If there is a deficiency of protein as in kidney disease when protein is constantly lost as albumin in the urine, the osmotic pressure of the plasma is lowered and excess fluid escapes into the tissues. This excess fluid in the tissues is called oedema. The viscosity of the blood also assists in the maintenance of the blood pressure. Albumin is thought to be formed in the liver and globulin is derived from the group of white blood cells called lymphocytes. Fibrinogen and prothrombin are produced in the liver and are both necessary for the mechanism of the clotting of blood. Plasma without fibrinogen is called serum; this can be seen as the yellow fluid which oozes from a cut after a clot has formed. Heparin is also formed in the liver and prevents blood clotting in the vessels.

Foodstuffs, in the form of glucose, amino acids, fatty acids and glycerol, are absorbed from the alimentary tract into the blood. They are the end-products of carbohydrate, protein and fat metabolism.

Urea, uric acid and creatinine are the *waste products* from protein metabolism. They are made in the liver and are carried by the blood for excretion by the kidneys.

Antibodies and antitoxins are complex protein substances which provide protection against infection and neutralize the poisonous bacterial toxins.

Enzymes are chemical substances produced by the body, which produce chemical changes in other substances without themselves entering into the reaction.

The Blood Cells

The cells are of three types: red blood cells (erythrocytes), white blood cells (leucocytes) and platelets (thrombocytes).

Red Cells. The red cells are minute disc-shaped bodies, concave on either side. They are very numerous, numbering about 5 000 000 per cubic millimetre of blood. They are very minute, having a diameter of 7.2 micrometres only (1 micrometre = 1/1000 millimetre; this can be abbreviated to μm). They have no nucleus, but contain a special protein known as *haemoglobin*. This is a pigment and is yellow in colour, though the massed effect of these numerous yellow bodies is to make the blood red. Haemoglobin contains a little iron, and this iron is essential to normal health, though the total amount in the whole body is said to be only sufficient to make a two-inch nail. Haemoglobin has a great attraction for oxygen. As the red cells pass through the lungs the haemoglobin combines with oxygen from the air (forming oxyhaemoglobin) and becomes bright in colour. This makes the oxygenated blood bright red. As the red cells pass through the tissues oxygen is given off from the blood and the haemoglobin becomes a dull colour (reduced haemoglobin) making the blood a dark purplish-red. The haemoglobin is measured in grams per 100 ml; the normal figure is 14 to 16 g per 100 ml.

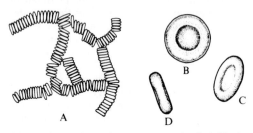

Fig. 89. Red cells. (A) In rouleaux, as seen in shed blood under the microscope; (B), (C), and (D) three views of a single cell.

The *function* of the red cells is therefore to carry oxygen to the tissues from the lungs and to carry away some carbon dioxide. This is their sole function, and is dependent on the amount of haemoglobin they contain. If there is a lack of haemoglobin, either because red cells are reduced in number or because each one does not contain the normal quantity of haemoglobin, the individual suffers from anaemia.

The red cells are *produced* in the red bone marrow of spongy bone, which is found in the extremities of long bones and in flat and irregular bones. In childhood the red bone marrow also extends throughout the shaft of the long bones, as children have greater need for the production of red cells.

Red cells pass through various stages of development in the bone marrow. Erythroblasts are large cells containing nuclei and a small quantity of haemoglobin. These develop into normoblasts which are smaller cells with more haemoglobin and smaller nuclei. The nucleus then disintegrates and disappears and the cytoplasm contains fine threads, at which stage the cells are called reticulocytes. Finally the threads disappear and the fully mature erythrocyte passes into the blood stream. In health almost all red cells in the blood should be erythrocytes, with only an occasional reticulocyte. Many factors are necessary for the normal formation of red blood cells.

Protein is required for the manufacture of protoplasm.

Iron is needed for the haemoglobin. Very little iron is excreted. As the red cells are broken down the iron is stored and used again, but a certain amount of iron must be taken in in the diet. A man requires about 10 mg of iron per day, women require about 15 mg to make good the menstrual loss and the depletion of iron reserves which occur during pregnancy, labour and lactation. Iron-containing foods are red meat, egg-yolk, green vegetables, peas, beans and lentils.

Vitamin B_{12} (cyanocobalamin) is necessary for the maturation of red blood cells and is usually found in adequate quantities in the diet in temperate climates. It can be absorbed from the small intestine only when it has been combined with the intrinsic factor which is secreted by the stomach. Together these two substances are known as the anti-anaemic factor (or haemopoietic factor), which is stored in the liver and passed to the bone marrow as necessary. Vitamin B_{12} is also known as the extrinsic factor.

Other factors which are necessary even though in small quantities are vitamin C, folic acid (one of the vitamin B complex), the hormone thyroxine and traces of copper and manganese.

Red blood cells live in the circulation for about 120 days after which they are ingested by the cells of the reticulo-endothelial system in the spleen and lymph nodes. Here the haemoglobin is broken down into its component parts, which are carried to the liver. The globin is returned to the protein stores or is excreted in the urine after further breaking down. The haem is further split into iron, which is stored and used again, and pigment, which is converted by the liver into bile pigments and is excreted in the faeces. Red cell production and breakdown usually proceed at the same rate so the number of cells remains constant.

White Cells. The white cells, or leucocytes, are larger than the red cells, measuring about 10 μm in diameter, and they are less numerous. There are 7–10 $\times$ 10^9 per litre of blood, though this number increases considerably to 30 $\times$ 10^9 per litre when infection is present in the body. This increase is known as leucocytosis. The leucocytes are of three different types:

1. Polymorphonuclear leucocytes are also known as granulocytes

because of the granular appearance of the cytoplasm. The nucleus gradually develops several lobes, hence the name (poly = many, morph = form). These cells make up about 75 per cent of the total white cells. They are made in the red bone marrow and survive for about 21 days. Granulocytes can be further classified according to their staining properties. Neutrophils (70 per cent) absorb both acid and alkaline dyes. They have the ability to ingest small particles, e.g. bacteria and cell debris. This power is called phagocytosis and they are sometimes known as phagocytes. They have amoeboid movement and can pass out of the blood stream through the capillary walls to accumulate where there is infection.

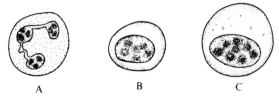

FIG. 90. White cells. (A) A polymorphonuclear leucocyte; (B) a lymphocyte; (C) a monocyte.

Eosinophils (four per cent) absorb acid dyes and stain red. An increase in their number occurs during allergic states such as asthma, and during infestation with worms.

Basophils (one per cent or less) absorb alkaline or basic dyes and stain blue. They contain heparin and histamine.

2. Lymphocytes make up about 20 per cent of the total white cell count. They are made in the lymph nodes and in the lymphatic tissue which is present in the spleen, the liver and other organs. They show some amoeboid movement but are not actively phagocytic. They are concerned in the production of antibodies.

3. Monocytes make up about five per cent of the total white cell count. They are the largest of the white blood cells and have a horse-shoe-shaped nucleus. They show both amoeboid movement and are phagocytic, and are part of the reticulo-endothelial system (see page 134).

Platelets. Platelets, or thrombocytes, are even smaller than red blood cells and are made in the bone marrow. There are about 250×10^9/litre of blood. They are necessary for the clotting of blood.

The Clotting of Blood

When the blood flows over a rough surface an enzyme called thrombokinase is released from the damaged tissue cells or damaged thrombocytes. In the presence of this enzyme, and when adequate

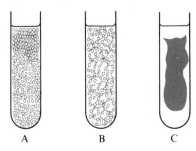

FIG. 91. Clotting of blood. (A) Blood shed; (B) fibres formed; (C) clot floating in serum.

amounts of calcium are also present, prothrombin, which is a protein normally present in the plasma, is converted into a new substance called thrombin. Thrombin is also an active enzyme which acts on fibrinogen, another of the normal plasma proteins, to produce an insoluble thread-like substance called fibrin. The fibrin threads entrap blood cells to form a clot. After some time the clot shrinks and serum is released (serum = plasma − fibrinogen).

Table 9

Stages in Clot Formation

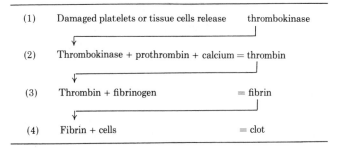

(1)   Damaged platelets or tissue cells release   thrombokinase

(2)   Thrombokinase + prothrombin + calcium = thrombin

(3)   Thrombin + fibrinogen   = fibrin

(4)   Fibrin + cells   = clot

Factors Affecting Clotting

Prothrombin is made in the liver and vitamin K is necessary for its manufacture. Vitamin K is present in green vegetables and is also manufactured in the intestines by bacterial action. It can be absorbed from the intestines into the blood only in the presence of bile. If bile is not present, as in some forms of jaundice, prothrombin may be lacking and the tendency to bleed is increased.

Heparin is a protein normally present in blood, which is formed in the liver and which prevents blood clotting in the vessels. It is called an anticoagulant.

The Functions of Blood

The uses of blood are:

1. To carry food to the tissues
2. To carry oxygen to the tissues in oxyhaemoglobin
3. To carry water to the tissues
4. To carry away waste products to the organs which excrete them
5. To fight bacterial infection through the white cells and antibodies
6. To provide the materials from which glands make their secretions
7. To distribute the secretions of ductless glands, and enzymes
8. To distribute heat evenly throughout the body, and so regulate the body temperature
9. To arrest haemorrhage through clotting

Blood Groups

Blood from one individual cannot always be safely mixed with that of another. This fact became evident with the introduction of blood transfusion, which at first sometimes cured but sometimes killed the patient. This was due to the fact that blood is of four basic groups. If blood of differing groups is mixed, the red corpuscles may become sticky and form clumps. This is termed *agglutination*, and is fatal as the clumps of red cells block the blood vessels and obstruct the circulation, and the kidneys are severely damaged by excretion of excessive quantities of pigment from destroyed red cells.

When a person requires a blood transfusion it is necessary first to find which group he belongs to and then to find a donor belonging to the same group. Blood groups are named according to the presence or absence of substances called *agglutinogens*, which are present in the red blood cells. There are two agglutinogens, called A and B. If A is present the blood group is called Group A, if B is present the blood group is called Group B, if both A and B are present the blood group is called Group AB, and if neither agglutinogen is present the group is called Group O. Having found to which group the patient belongs (Table 10) and found blood donated by a person belonging

Table 10

Blood group	Agglutinogen in red cells	Agglutinin in plasma	Transfusion possible
A	A	Anti-B	Groups A and O
B	B	Anti-A	Groups B and O
AB	A and B	Neither	Any group
O	None	Anti-A and anti-B	Group O only

to the same basic group, a sample of red blood cells from the donor's blood is mixed with some plasma from the patient who is to receive the transfusion (now called the recipient). This is because plasma contains substances called *agglutinins* which cause agglutination of red cells if incompatible blood groups are mixed. Agglutinins are called anti-A and anti-B, and plasma contains all agglutinins which will not affect its own red cells. Therefore the plasma of Group A contains anti-B agglutinin, the plasma of Group B contains anti-A agglutinin, the plasma of Group AB contains no agglutinins, and the plasma of Group O contains both anti-A and anti-B agglutinins. When the donor's red cells are mixed with the recipient's plasma in the laboratory, it can be seen with the help of a microscope whether agglutination occurs. It will be noticed that Group AB has no agglutinins in the plasma and therefore cannot cause any red cells to agglutinate; this means that a patient with blood belonging to this group will probably be able to receive blood from any other group and the group is therefore known as the universal recipient. Group O contains no agglutinogens in the red cells and therefore they cannot be made to agglutinate by the agglutinins in any plasma. Blood belonging to this group can therefore probably be given to a patient belonging to any group and it is known as the universal donor. In practice the compatibility of blood is always checked very carefully before it is given to a patient.

Rhesus Factor

In addition to ABO grouping there is an additional factor present in the blood of about 85 per cent of the population. It is an agglutinogen called the Rhesus factor. Those who possess the factor are called Rhesus positive (Rh+), the remaining 15 per cent are called Rhesus negative (Rh−). If an Rh− person receives the blood of an Rh+ donor the agglutinogen stimulates the production of anti-Rh agglutinins called anti-D. If a second Rh+ transfusion were given later the transfused cells would be agglutinated and destroyed (haemolysed) with serious or fatal results to the recipient. This factor can also cause difficulty during pregnancy. If an Rh− mother is carrying an Rh+ fetus the mother may begin to produce anti-Rh agglutinins which may then destroy the baby's red cells. The baby may overcome this spontaneously, or may require exchange transfusion.

Immunity

When a foreign body is introduced into the body the immediate response is the production of a substance which will react with it and render it harmless. The foreign protein is called an antigen and substances produced in response to the antigen are called *antibodies*. Antigens can be any foreign protein, but common antigens are micro-organisms, some drugs (penicillin is an example), animal and vegetable proteins, including pollens, and foreign tissues such as transplanted organs. The reaction that occurs may be called an antigen–antibody reaction. When it occurs as a response to micro-organisms it is called immunity.

Immunity is a useful defence against infection. Each type of micro-organism which enters the body acts as an antigen and stimulates the production of specific antibody which destroys that antigen and no other. The first time a person comes in contact with the virus which causes measles, antibody is produced which enables the body to overcome the infection and which remains in the blood ready to prevent any further infection by the same type of virus. Immunity may be active, when the cells of the body make the antibody, or may be passive, when the antibody has been produced in the cells of another person.

Active immunity may be achieved in several ways. *Active natural immunity* is gained by having the disease, after which the antibody remains in the blood ready to prevent another attack of the same disease. This type of immunity is also developed by what are called sub-clinical infections, in which the body is exposed to small numbers of micro-organisms insufficient in number to give rise to any definite symptoms, but which are sufficient to stimulate antibody production. *Active artificial immunity* is given to children and travellers in order to prevent their getting diseases which would be serious or fatal. An injection of killed micro-organisms, or live ones which have been rendered harmless, is given and the body responds by producing antibodies. In this way active immunity is built up. Harmless toxins are also used to give this type of immunity. Toxins are chemical poisons produced by micro-organisms and when rendered harmless they also act as antigens. Harmless micro-organisms are called vaccines and harmless toxins are called toxoids. Many diseases are prevented by active artificial immunity; some of the most common are whooping cough, diphtheria, measles, small-pox, poliomyelitis and tuberculosis.

Table 11

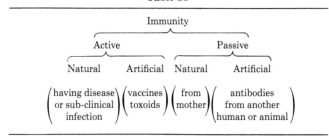

Passive natural immunity is gained by the baby before birth as the antibodies are passed from the mother to the fetus. *Passive artificial immunity* is useful in the prevention of disease or in its treatment. The antibodies are produced in another human or in an animal and are injected into the person at risk. Passive immunity is always short-lived as the antibodies are destroyed after a short time.

Antigen—antibody reactions normally occur in the bloodstream and the debris is carried away by the reticulo-endothelial system. When the immune reaction occurs in the tissues themselves the cells are damaged or destroyed by its side effects and this is known as *allergy*. Allergic reactions are often due to protein-like substances called *allergens*. The allergic reaction in the tissues releases *histamine* which causes redness and swelling in the skin, as in urticaria, and the outpouring of fluid as in hayfever. There may also be constriction of unstriped muscle in the respiratory tract causing asthma.

The word *autoimmunity* describes the situation in which the body makes antibodies against some of its own cells. Many diseases are thought to be autoimmune in origin, two of the best known being rheumatoid arthritis and rheumatic fever.

Reticulo-endothelial System

Monocytes have already been described as being one type of white blood cell which are actively phagocytic and mention was made of their being part of the *reticulo-endothelial system*. This system is widespread throughout the body and its cells have the ability to move about which, combined with their phagocytic powers, makes them one of the most important defences of the body against micro-organisms. They are also concerned with the production of antibodies.

The cells of the system occur in the following parts of the body:

1. Connective tissue, where they are called *histiocytes*
2. Blood, where they are called *monocytes*
3. Lining blood vessels of the bone marrow, spleen and liver and parts of the suprarenal gland and the anterior lobe of the hypophysis
4. In the lymph nodes, lymph follicles of the small intestine and the tonsils
5. The meninges where they are called *meningiocytes*.

13 The Heart and Blood Vessels

The heart is a hollow, muscular, cone-shaped organ; it lies between the lungs in a block of tissue called the *mediastinum*, behind the body of the sternum with two-thirds of its bulk on the left side. The circular *base* of the cone is directed upwards and to the right and the *apex* points downwards, forwards and to the left. It is usually on the

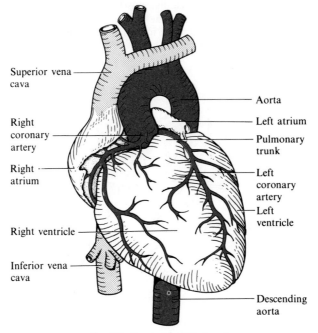

FIG. 92. Front view of the heart.

level of the fifth intercostal space about 9 cm from the midline. The heart measures about 12 cm from base to apex, about 9 cm in width and is about 6 cm thick.

General Structure of the Heart

The heart is divided from base to apex by a muscular partition called the *septum*. The two sides of the heart have no communication with each other in health. Additionally each side is subdivided into an upper and lower chamber. The upper chamber on each side, the *atrium*, is smaller and is a receiving chamber into which the blood flows through veins. The lower chamber, the *ventricle*, is the discharging chamber from which the blood is driven into the arteries. Each atrium communicates with the ventricle below it on the same side of the heart through an opening, guarded by a valve called the *atrio-ventricular valve*.

The heart is composed of cardiac muscle, the *myocardium*, on the action of which the circulation of the blood depends. The myocardium varies in thickness, being thickest in the left ventricle, thinner in the right ventricle and thinnest in the atria.

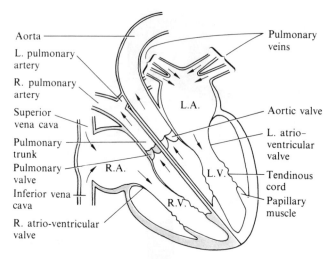

FIG. 93. Diagram showing the flow of blood through the heart. L.A., left atrium; L.V., left ventricle; R.A., right atrium; R.V., right ventricle.

The atria and ventricles are lined with a thin, smooth, glistening membrane called the *endocardium* which consists of a single layer of endothelial cells and which is continuous with the valves and with the lining of the blood vessels.

The *pericardium* covers the heart and the roots of the great vessels and has two layers. The outer layer or *fibrous pericardium* is securely anchored to the diaphragm, the outer coats of the great vessels and the posterior surface of the sternum and therefore maintains the heart in its position in the chest. Because of its fibrous nature it also prevents over-distension of the heart. The inner layer, the *serous pericardium*, lines the fibrous pericardium and is invaginated by the heart; it therefore also has two layers. The inner layer is known as the visceral portion, or *epicardium*, and it is reflected back to form the outer or parietal layer. The layers are normally all in close contact and are moistened by fluid which exudes from the serous membrane; this prevents any friction as the heart continually contracts and relaxes. In inflammatory conditions, such as pericarditis, the amount of fluid within the pericardium may embarrass the action of the heart and may be aspirated.

Valves of the Heart. The heart is provided with *valves* to prevent the blood from flowing in the wrong direction. There are four main valves.

The *right atrio-ventricular (or tricuspid) valve* lies between the right atrium and the right ventricle. It consists of three triangular flaps or cusps, each consisting of a double layer of endocardium strengthened with fibrous tissue. The under surface of the cusps gives attachment to a number of fine, tendinous cords, called *chordae tendinae*, which originate in the *papillary muscles* in the wall of the ventricle. When the ventricle contracts the blood is pushed back towards the atrio-ventricular opening but is prevented from entering the atrium by the cusps of the valve, which close because of the increased pressure in the ventricle. The contraction of the papillary muscles exerts tension on the chordae tendinae which prevent the valve cusps from being carried into the right atrium.

The *left atrio-ventricular valve* is also called the mitral valve because it has only two cusps. The structure is similar to that of the right atrio-ventricular valve. It prevents the backflow of blood into the left atrium during contraction of the left ventricle.

The *aortic valve* consists of three cusps which surround the entrance into the aorta from the left ventricle. The cusps are half-

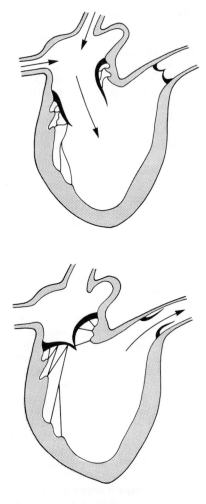

Fig. 94. The action of the valves of the heart. Top: the position of the valves during contraction of the atrium and relaxation of the ventricle. Bottom: the position of the valves during contraction of the ventricle and relaxation of the atrium.

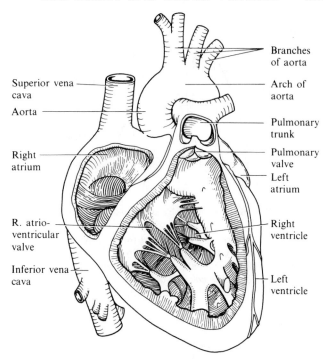

FIG. 95. Inside the right side of the heart.

moon shaped and are fixed by their curved edges to the wall of the aorta, the straight edge being free so that pockets are formed facing into the aorta. As the blood flows from the left ventricle into the aorta the cusps lie flat against the vessel wall; as the ventricle relaxes the pockets fill with blood and bulge out, meeting in the centre and blocking the opening completely thus preventing blood flowing back into the ventricle. The *coronary arteries*, which supply the heart muscle with oxygenated blood, arise from the aorta just above the attached edges of the cusps of the aortic valve.

The *pulmonary valve* guards the opening from the right ventricle into the pulmonary trunk. It is similar in structure and action to the aortic valve.

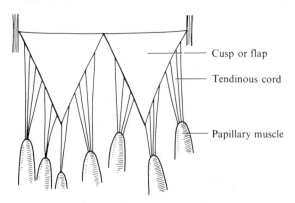

Fig. 96. The mitral valve laid open (diagrammatic).

The blood which returns to the heart from the myocardium passes through the *coronary sinus* and pours directly into the right atrium. The opening of the coronary sinus is protected by a thin, semi-circular valve called the *valve of the coronary sinus* which prevents backflow of blood into the sinus during contraction of the right

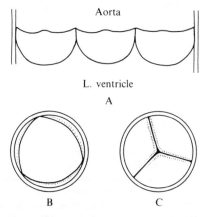

Fig. 97. The aortic valve (diagrammatic). (A) Laid open, (B) seen from above, and (C) closed.

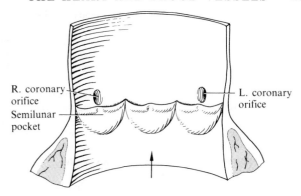

R. coronary orifice

L. coronary orifice

Semilunar pocket

FIG. 98. The aortic valve.

atrium. There is also an imperfect valve guarding the opening from the inferior vena cava into the right atrium; it is called the *valve of the inferior vena cava*.

The blood supply to the heart muscle is by the *right and left coronary arteries* which are the first branches from the aorta. There are a number of places within the heart muscle where these two arteries join, or anastomose, but most of the blood returns from the myocardium in veins which empty into the *coronary sinus*.

The Function of the Heart

The heart is a pump whose purpose is to drive the blood into and through the arteries but the right and left sides of the heart function quite separately from one another.

Blood from all parts of the body is returned to the *right atrium* through the two large veins, the *superior* and *inferior venae cavae*. When it is full the right atrium contracts and drives the blood through the right atrio-ventricular valve into the *right ventricle* which in turn contracts sending the blood through the *pulmonary valve* and into the *pulmonary trunk*. The pulmonary trunk divides into *right* and *left pulmonary arteries* which carry the blood to the lungs where the gaseous exchange occurs. The blood is finally collected up into four *pulmonary veins* which return the blood to the *left atrium*. When it is full the left atrium contracts, simultaneously with the right atrium, and the blood is driven through the left atrio-ventricular valve into the *left ventricle*. This chamber contracts, simultaneously with the right ventricle, and sends the blood into the *aorta*, the main artery of the body.

The heart contracts about seventy to eighty times each minute throughout life, though the rate varies with age, emotion, exercise and other influences. Each beat is a cycle of events which lasts about 0·8 second.

Blood pours into the atria from the great veins until both are full and they then contract simultaneously, emptying their contents into the ventricles. Atrial contraction lasts about 0·1 second. Rising pressure in the ventricle as blood enters forces the atrio-ventricular valves to close and causes the first heart sound, which can be heard through a stethoscope placed over the apex of the heart and has been likened to the word 'lûbb'. Ventricular contraction follows, lasting 0·3 second, and forcing open the pulmonary and aortic valves. Blood is forced into the aorta and pulmonary trunk and as the ventricles relax the pressure in the great vessels forces the aortic and pulmonary valves to close and causes the second heart sound, which can be heard best over the second right rib and has been likened to the word 'dŭp' because it is a sharper sound. During ventricular contraction the atria are relaxed. Following ventricular con-

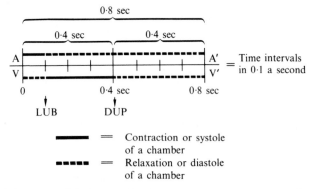

FIG. 99. Graph to show the sequence of events in the cardiac cycle. A–A': the contraction and relaxation of the atrium; V–V': the contraction and relaxation of the ventricle. The cycle occupies 0·8 sec of which 0·4 sec are occupied by the contraction of both atria and ventricles, and the remaining 0·4 sec by relaxation of all chambers of the heart. (Note 0·8 sec is the time of one beat if the pulse rate is 75 per minute.)

traction the whole heart is relaxed for approximately 0·4 second. The word for contraction is *systole* and for relaxation, *diastole*.

The Conducting Mechanism of the Heart

Cardiac muscle, unlike skeletal muscle, has the property of being able to contract rhythmically independent of any nerve supply. The autonomic nervous system modifies the speed of the heart beat but the heart could be removed from an animal and continue to beat. The impulse to contract arises in an area of specialized tissue near the entry of the superior vena cava into the right atrium—called the *sinu-atrial node*—the pacemaker of the heart. The impulse then passes across both atria in more or less concentric rings, made possible by the fact that cardiac muscle fibres are branching, to the *atrio-ventricular node* situated in the septum near the junction of the atria and ventricles. After a slight pause the impulse spreads down the *atrio-ventricular bundle* which is in two strands, one in the right and one in the left ventricle. The strands break up into special fibres, called *Purkinje fibres*, from which branches pass beneath the endocardium to all parts of the ventricles.

The sinu-atrial node has the fastest rhythm, about 70 to 74 beats per minute and it imposes this rate on the other areas of conduction. The ventricles are able to contract independently of the atria and they will do this if the conducting mechanism is affected by disease, but they will contract much more slowly, about 40 beats per minute. This condition is known as *heart block* and may be serious as the tissues are unlikely to receive an adequate blood supply. In other cases some impulses pass down under the atrio-ventricular bundles but some do not so the ventricles contract once to every two or three atrial contractions; the condition is known as partial heart block.

Nervous Control of the Heart

The heart is innervated by the autonomic nervous system. The *vagus nerve* (tenth cranial nerve) slows the heart rate and causes decreased power of contraction by conveying impulses to the sinu-atrial node. The *sympathetic nerves* speed the heart rate and increase the force of contraction. (See also Chapter 22.) This dual innervation of the heart is coordinated by the cardiac centre in the medulla oblongata in the brain.

The heart rate is also controlled reflexly by two sets of receptors. *Pressure receptors*, or baroreceptors, are sensitive to changes in blood pressure. They are found in the carotid arteries and the arch of the aorta. If the blood pressure increases the sympathetic nerves are depressed and the heart rate slows, thus helping to lower the blood pressure. *Chemoreceptors* are sensitive to the amount of oxygen and carbon dioxide present in the blood. They are found in the neck near the carotid arteries and near the aorta and are sensitive to lack of oxygen. Impulses are conveyed to the cardiac centre and the heart rate is accelerated to increase the blood supply, and therefore the oxygen supply, to the tissues.

In health the heart rate varies considerably:

1. Rest slows the heart rate and exercise quickens it.
2. Increasing age slows the heart rate. Infants have a pulse rate of 120 to 140 at birth and it slows throughout life and into old age.
3. Women have a slightly faster pulse rate than men.
4. Emotion and excitement speed the pulse rate.

In disease, conditions such as fever, haemorrhage, shock and hyperthyroidism speed the heart rate, while increased pressure on the brain and heart block slow it. Drugs can also be used to speed or slow the pulse rate.

The Blood Vessels

Contraction of the ventricles sends the blood to all parts of the body through a complicated series of tubes called *arteries* which branch into tiny vessels called *arterioles*. These in turn are continuous with a network of microscopic vessels named *capillaries*. Blood is then collected into tiny vessels called *venules* which unite to form *veins*. These join one another forming large veins which finally return the blood to the heart.

Structure of the Blood Vessels

Arteries are thick-walled vessels and with one exception they carry oxygenated blood. The exception is the pulmonary trunk, which divides into two pulmonary arteries, and which carries deoxygenated blood from the right ventricle to the lungs. All arteries have three coats:

1. The outer coat, or *tunica adventitia*, which consists of collagenous and elastic fibres.
2. The middle coat, or *tunica media*, which consists mainly of unstriped muscle with elastic fibres and some collagen fibres.

3. The lining, or *tunica intima*, which consists of a layer of endothelial cells and provides a smooth surface over which the blood can flow without clotting.

All parts of the body must have a blood supply and the arteries are no exception. Very tiny blood vessels bring a blood supply to the arterial walls and equally small venules return the blood to the veins. Lymph vessels and nerve fibres are also present.

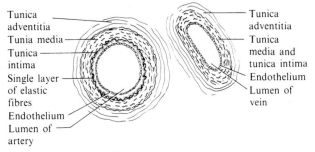

FIG. 100. Structure of artery and vein.

Arterioles have the same three structures as arteries, but the intima and media are thin and the adventitia is relatively thicker than the adventitia of an artery. There are also more muscle fibres and less elastic fibres.

Capillaries form a network between arterioles and venules. They consist of a single layer of endothelial cells, similar to those which line all other blood vessels.

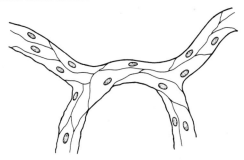

FIG. 101. Capillary network showing the wall of simple endothelium.

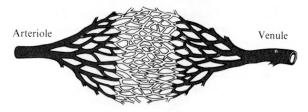

Network of capillaries

Fig. 102. Diagram of a capillary network.

Venules and *veins* both have three coats as do arteries but the middle coat of a vein is much thinner than that of an artery. Many veins are provided with *valves* to prevent blood flowing in the wrong direction. Each valve is formed of a double layer of endothelium strengthened by connective tissue and elastic fibres. The valve cusps are semi-lunar and are attached by their convex edges to the wall of the vein. Veins, like arteries, have their own blood supply and nerve fibres though the latter are not so numerous.

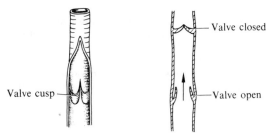

Fig. 103. Valves in the veins.

The Mechanism of Circulation

The Pulse

Blood leaving the left ventricle of the heart is rich in oxygen and bright red in colour. It is pumped into the aorta by contraction of the left ventricle which initiates an area of increased pressure which travels along the arteries rather like a wave. When the blood is pumped out of the left ventricle the aorta is already full, so it must distend in order to accommodate the additional blood. As the left ventricle relaxes the aortic valve closes and the elastic aorta recoils to its original diameter. This recoil of the aorta is very important because it is the mechanism whereby the blood is continuously driven round the body even when the ventricle is relaxed. The distension and recoil of the aorta sets up a wave of distension and recoil called the *pulse* which travels along all the large arteries and which can be felt with the fingers wherever an artery can be compressed gently against a bone. Since the heart beat produces the pulsation the rate and character of the beat can be judged by taking the pulse.

Blood Pressure

The blood pressure is the force which the blood exerts on the walls of the blood vessels. It varies in the different blood vessels and also with the heart beat. The pressure is greatest in the large arteries leaving the heart, and gradually falls in the arterioles until, when it reaches the capillaries, it is so slight that the least pressure from without will obliterate these vessels and drive the blood out of them. This can be seen by pressing lightly on the nail or letting a piece of glass rest lightly on the skin. (It is for this reason that it is so important to change frequently the position of a patient confined to bed, as the tissue carrying the body's weight has little blood circulating through it.) In the veins the pressure is lower still until ultimately in the big veins approaching the heart there is suction, i.e. a negative pressure instead of a positive one, on account of the suction exerted by the heart as its chambers relax.

The pressure in the large arteries varies with the heart beat. It is highest when the ventricle contracts (this is called systolic pressure) and lowest when the ventricle relaxes (this is called diastolic pressure).

Measurement of Blood Pressure. The pressure of the blood is measured by the height of the column of mercury which it will support. The height is calculated in millimetres. The normal arterial pressure is 110 to 120 millimetres systolic pressure and 65 to 75 millimetres diastolic pressure. The apparatus for measuring blood pressure is called a sphygmomanometer. There are different types, but the most common consists of a hollow rubber cuff encased in material which is fixed evenly round the limb without constricting it in any way. To this is attached a small hand pump, by which air can

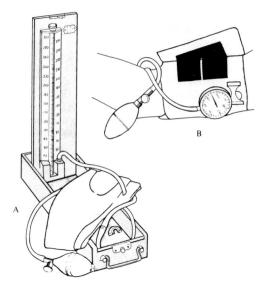

FIG. 104. (A) Diagnostic sphygmomanometer, and (B) an aneroid
sphygmomanometer.

be pumped into the cuff, expanding it and compressing the limb
and the blood vessels it contains. The air cushion is also linked to a
manometer containing mercury, so that the pressure of the air within
it can be read against a graduated scale.

To take the blood pressure the air cuff is fixed round the arm and
the pulse at the wrist is felt with one hand while the other is used to
inflate the cuff to a pressure above the point at which the radial
pulse can no longer be felt. A stethoscope is placed over the brachial
artery pulse in the cubital fossa and the pressure in the cuff is slowly
lowered by means of the release valve. As the pressure falls no sounds
are heard until the systolic blood pressure is reached, at which point
a tapping sound is heard in the stethoscope. At this point the level of
mercury in the manometer must be noted. As the pressure in the cuff
is further reduced the sounds become louder, until the diastolic blood
pressure is reached, at which point the sound is altered in character
and becomes muffled. A little further decrease in cuff pressure

causes the sound to disappear completely. The diastolic blood pressure is noted as the point at which the character of the sound changes.

Arterial Pressure

The arterial pressure is maintained by:

1. The cardiac output
2. The peripheral resistance
3. The total blood volume
4. The viscosity of the blood
5. The elasticity of the arterial walls

The *cardiac output* is the amount of blood pumped out with each ventricular contraction. When the left ventricle contracts about 70 ml of blood is pushed into the aorta, which is already full, causing it to distend.

The *peripheral resistance* is the resistance offered by the small blood vessels, particularly the arterioles, to the flow of blood. This resistance prevents blood flowing too quickly into the capillaries and in this way helps to maintain the blood pressure in the arteries. The lumen of the arterioles can be altered by the action of vasomotor nerves and by adrenaline and noradrenaline from the adrenal glands. If the lumen is narrowed the resistance to blood flow is increased and arterial blood pressure is raised. If the lumen is enlarged more blood will pass through quickly to the capillaries and the blood pressure will be lowered.

The *blood volume* is the total amount of circulating blood. If this is reduced by the loss of whole blood, as in haemorrhage, or by loss of fluid from the circulation, as in shock, burns or dehydration, the blood pressure will be lowered.

The *viscosity of the blood* is its stickiness. Blood is a viscous fluid which offers two or three times as much resistance as plain water. The viscosity depends partly on the plasma, particularly the plasma proteins, and partly on the number of red cells. The viscosity of the blood will be reduced if large quantities of intravenous normal saline are given, and lowered viscosity is associated with lowered blood pressure.

The *elasticity of the arterial walls* allows distension of the aorta when the ventricle contracts and elastic recoil as the ventricle relaxes. This recoil pushes the blood on and maintains the diastolic pressure of the blood. Distension and recoil occur throughout the arterial system. Elasticity may be reduced in atheroma which is a degenerative disease of the arteries. In this condition the blood pressure will be raised because of loss of elasticity.

High Blood Pressure. A high blood pressure is a systolic pressure of 150 to 180 mm and over. The diastolic pressure is also raised as a general rule, and a high diastolic pressure, e.g. 90 to 120 mm or more, is injurious, as it is a strain on the heart. A high blood pressure is dangerous in that the raised pressure may cause rupture of a blood vessel. Rupture is particularly likely to occur in the brain and is one of the causes of cerebrovascular accident or stroke. There is also a strain on the heart which may result in heart failure.

Low Blood Pressure. A low blood pressure is a systolic pressure of 100 mm or less. It occurs in cases of haemorrhage, shock and collapse, heart failure, and in disease of the suprarenal glands. The danger lies in an insufficient supply of blood to the vital centres of the brain. The treatment is therefore:

1. To place the patient flat with the foot of the bed raised if necessary, to send blood by gravity to the vital centres of the brain.
2. To give heart stimulants, e.g. nikethamide, adrenaline or noradrenaline, and circulatory stimulants such as adrenaline and noradrenaline to contract the blood vessels.
3. To give fluids such as saline by intravenous, subcutaneous or rectal infusion, or blood or plasma transfusion, to increase the fluid in circulation.

As already mentioned, blood vessels have their own nervous supply. Arterioles are supplied with *vasomotor nerves* through which they can be dilated or contracted according to the needs of the tissues. The muscular walls of the arterioles are also affected by hormones from the suprarenal glands, particularly adrenaline and noradrenaline, which cause contraction of the muscle fibres and consequent narrowing of the lumen of the arteriole. Arterioles do not pulsate but can dilate to carry more blood to an organ which is working and can contract so that they carry less blood when an organ is resting. In this way arterioles control the distribution of blood to the various organs of the body and are important in the maintenance of blood pressure as their contraction offers resistance to the outflow of blood from the large elastic arteries.

Capillaries receive blood from the arterioles and pass it to the venules. Because their walls are only one cell in thickness oxygen, water and food substances in solution are able to pass from the blood through the walls to supply the tissue cells and waste products can pass back from the tissues to be carried away by the blood.

Capillaries form a network with considerable numbers of them forming anastomoses so that varying amounts of blood may be brought to the part as required.

Venous Return. Blood is collected up from the capillary network into venules and then into veins. By the time it reaches the veins it has given up most of its oxygen and is a dark purplish colour. Valves are necessary in the veins to prevent the blood flowing in the wrong direction.

Venous return depends on three factors:

1. Suction as the atria relax.
2. Suction during inspiration, which draws blood towards the heart as well as drawing air into the lungs.
3. Pressure on the thin-walled veins by contraction of muscles. As the lumen of the vein is reduced in size blood would be forced in both directions were it not for the presence of valves which allow the blood to flow towards the heart only.

The force of suction is the most important factor in the maintenance of venous pressure. If the force is lessened, as in heart failure, venous return is impaired and congestion occurs.

14 The Circulation

The circulation of the blood is divided into three main parts:

1. The systemic circulation
2. The pulmonary circulation
3. The portal circulation

The Systemic Circulation

The vessels which carry the blood from the left ventricle through the body generally to the right atrium are those which constitute the systemic circulation. Arteries may subdivide into several branches at the same point or several branches may be given off in succession. Arteries do not always end in capillaries but may unite with one another, forming *anastomoses*. Examples are found in the brain where two vertebral arteries anastomose to form the basilar artery and two anterior cerebral arteries are connected by the anterior communicating artery (see Fig. 106). An enlarged anastomosis may provide a *collateral circulation* if a vessel is occluded by accident or disease; sudden occlusion may be followed by death of the tissue supplied by the vessel whereas gradual occlusion may allow dilation of the anastomosis and adequate nourishment of the tissue. Some arteries have no anastomosis with other arteries and these are called *end-arteries*. Occlusion of an end-artery will cause death (necrosis) of the tissue supplied by the vessel; an example is the central artery of the retina, following occlusion of which permanent blindness will occur.

The Arteries

Arteries are named after the bones in the limbs or from the organ they supply. The most important thing about them is their position; where they can be compressed against a bone the pulse may be felt or pressure may be applied as a first aid measure in haemorrhage but at the same time it is important to ensure that there is no undue pressure being exerted on them by splints or other equipment. They are usually found on the inner side of a limb where there is only one bone, or between the bones where there are two, and at a joint they would cross the flexor surface. These positions are the safest because the vessels are less likely to be exposed to injury and pressure.

Aorta. The aorta is the main artery of those which carry oxygenated blood to the tissues of the body. It arises from the upper part of the left ventricle, passes upwards and to the right—the *ascending aorta*—and then arches backwards to the left—*the arch of the aorta*—and passes down through the thorax on the left side of the spine—*the descending thoracic aorta*. It enters the abdominal cavity through an opening in the diaphragm called the *aortic hiatus* and is then called the *abdominal aorta*. It ends at the lower border of the fourth lumbar vertebra by dividing into the right and left common iliac arteries.

The *ascending aorta* has two branches, the right and left coronary arteries which arise immediately above the cusps of the aortic valve. They supply the heart wall.

The *arch of the aorta* has three branches which arise from the top of the arch.

1. The *brachiocephalic trunk* which divides into the *right subclavian artery* and the *right common carotid artery*
2. The *left common carotid artery*
3. The *left subclavian artery*

The *common carotid arteries* supply the head and neck. Each divides into two at the level of the thyroid cartilage to form the external and internal carotid arteries. The *external carotid artery* supplies the outer parts of the face and scalp and has many branches such as the facial, temporal, occipital and maxillary branches. The *internal carotid artery* supplies a large part of the cerebrum, the eyes and the nose and forehead. At the point where the common carotid artery divides there is a dilated area called the *carotid sinus* where a large number of sensory nerve endings, from the glossopharyngeal (9th cranial) nerve, are situated. The sinus reacts to changes in arterial blood pressure and assists in making appropriate changes to return it to normal. A small reddish-brown structure behind the division of the common carotid artery is called the *carotid body*; it acts as a *chemoreceptor* (see page 144).

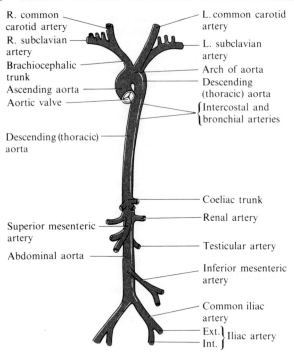

R. common carotid artery
R. subclavian artery
Brachiocephalic trunk
Ascending aorta
Aortic valve
Descending (thoracic) aorta

L. common carotid artery
L. subclavian artery
Arch of aorta
Descending (thoracic) aorta
{ Intercostal and bronchial arteries

Coeliac trunk
Renal artery
Superior mesenteric artery
Testicular artery
Abdominal aorta
Inferior mesenteric artery
Common iliac artery
Ext. } Iliac artery
Int. }

FIG. 105. The arch of the aorta and branches.

The *vertebral arteries* arise from the first part of the subclavian arteries and pass upwards in the foramina of the transverse processes of the cervical vertebrae; they enter the skull through the foramen magnum (see Fig. 47) and join together to form the *basilar artery*.

An anastomosis named the *circulus arteriosus* connects the vertebral arteries and the two internal carotid arteries. This circle is situated at the base of the brain. In front the two *anterior cerebral arteries* are joined by the *anterior communicating artery*. Behind, the basilar artery, formed by the junction of the two vertebral arteries, divides into two *posterior cerebral arteries*, each of which is

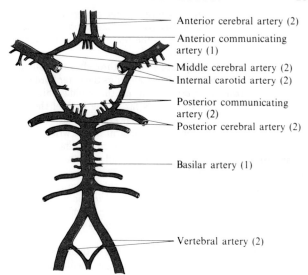

Fig. 106. The arterial circle of the cerebrum (circle of Willis).

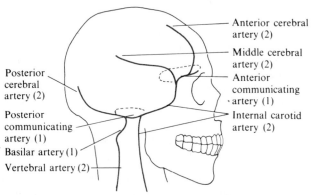

Fig. 107. Blood supply to the brain (side view). Broken lines indicate communication with the opposite side of the brain.

joined to the internal carotid arteries by the *posterior communicating artery*. This anastomosis increases the likelihood of an adequate blood supply being maintained to the brain if one of the vessels is injured or occluded.

Blood is supplied to the upper limb through one main artery which is called the *subclavian artery* as far as the outer border of the first rib, the *axillary artery* as far as the middle third of the humerus and the *brachial artery* as far as the back of the radius, where it divides into the radial and ulnar arteries. The *radial artery* passes along the radial side of the forearm to the wrist, where its pulsation can be felt quite easily, and then crosses into the palm of the hand to form the *deep palmar arch* and to unite with the deep branch of the ulnar artery.

The *ulnar artery* runs down the inner side of the forearm to the wrist which it crosses to form the *superficial palmar arch*. The palmar arches supply the digits and each anastomoses freely with the other so that injury to one allows blood to be brought to the part by other vessels.

The *descending thoracic aorta* is situated in the mediastinum; it supplies the pericardium, bronchi, oesophagus, mediastinum, intercostal muscles and breasts through branches which are named after the part they supply.

The *abdominal aorta* begins at the aortic hiatus of the diaphragm, about the level of the last thoracic vertebra. It has many large branch arteries arising from it to supply the abdominal organs and therefore it decreases rapidly in size.

1. The *phrenic arteries* supply the diaphragm.

2. The *coeliac trunk* arises just below the aortic opening in the diaphragm and divides into three branches: (a) the *left gastric artery* which supplies the stomach and gives off two or three branches which ascend through the oesophageal opening in the diaphragm and anastomose with the oesophageal arteries; (b) the *hepatic artery* which supplies the liver and gives off branches to the duodenum and bile duct as well as supplying the *right gastric artery*; (c) the *splenic artery* which divides into many branches to supply the very vascular spleen and also gives off numerous small vessels to supply the pancreas.

3. The *superior mesenteric artery* supplies all the small intestine, except part of the duodenum, and the first part of the large intestine.

4. The *middle suprarenal arteries* arise, one on each side of the

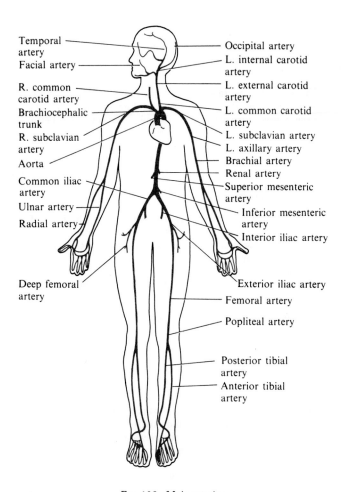

Temporal artery

Facial artery

R. common carotid artery

Brachiocephalic trunk

R. subclavian artery

Aorta

Common iliac artery

Ulnar artery

Radial artery

Deep femoral artery

Occipital artery

L. internal carotid artery

L. external carotid artery

L. common carotid artery

L. subclavian artery

L. axillary artery

Brachial artery

Renal artery

Superior mesenteric artery

Inferior mesenteric artery

Interior iliac artery

Exterior iliac artery

Femoral artery

Popliteal artery

Posterior tibial artery

Anterior tibial artery

FIG. 108. Main arteries.

aorta, opposite the superior mesenteric artery and supply the supra-renal glands.

5. The *renal arteries* supply the kidneys. The left artery is a little higher than the right because of the positions of the respective kidneys.

6. The *ovarian arteries* in the female and the *testicular arteries* in the male supply the organs after which they are named.

7. The *inferior mesenteric artery* supplies the remainder of the large intestine including the sigmoid colon and rectum.

The abdominal aorta divides into two *common iliac arteries* and these again divide, at the level of the last lumbar intervertebral disc, into the *internal iliac artery* which supplies the pelvis, the perineum and the gluteal region and the *external iliac artery* which supplies the lower limb.

Blood is supplied to the lower limb through the *external iliac artery* which crosses the groin and enters the thigh as the *femoral artery*; this in turn becomes the *popliteal artery* at the lower one-third of the thigh. The *popliteal artery* divides into:

1. The *anterior tibial artery*, which runs down the front of the leg on the anterior surface of the interosseous membrane, crosses the front of the ankle joint and supplies the front of the foot as the *dorsalis pedis* artery.

2. The *posterior tibial artery* which runs down the back of the leg to the back of the ankle joint and to the sole of the foot to become the *plantar arch*.

The arteries around the ankle joint anastomose freely with one another to form networks of vessels.

The Veins

The veins, which return venous blood to the heart, are either superficial veins which vary considerably in position, or deep veins which usually accompany arteries. Arterial pulsation is one of the factors which aids venous return. Systemic veins are more variable than corresponding arteries and anastomoses occur more frequently. In some areas, such as the pelvis and around the vertebral column, the veins form extensive anastomoses and often do not have valves.

The veins can be described in two groups:

1. The veins of the head, neck, upper limbs and thorax which all end in the superior vena cava.

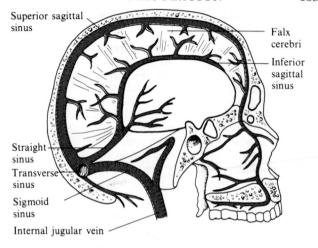

Superior sagittal sinus

Falx cerebri

Inferior sagittal sinus

Straight sinus

Transverse sinus

Sigmoid sinus

Internal jugular vein

FIG. 109. Venous sinuses within the cranium.

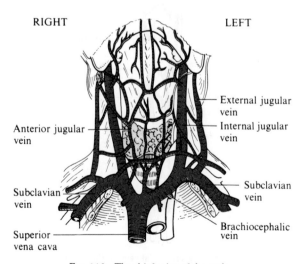

RIGHT

LEFT

External jugular vein

Internal jugular vein

Anterior jugular vein

Subclavian vein

Subclavian vein

Superior vena cava

Brachiocephalic vein

FIG. 110. The chief veins of the neck.

2. The veins of the lower limbs, abdomen and pelvis which all end in the inferior vena cava.

The blood from the brain is collected into vessels which are situated between the two layers of the dura mater and are called *venous sinuses*. These in turn empty into the *internal jugular veins* along with blood from the superficial parts of the face and from the neck.

The *external jugular veins* receive blood from the exterior of the cranium and from the deep parts of the face. At the root of the neck the internal jugular veins join with the *subclavian veins* to form the *brachiocephalic veins* which in turn unite to form the superior vena cava through which blood is poured into the right atrium.

The veins of the upper limbs are in two groups: superficial and deep.

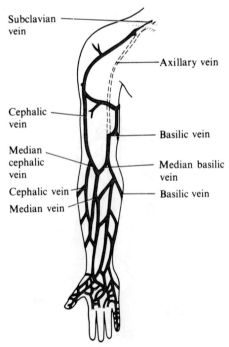

FIG. 111. The superficial veins of the forearm.

The *superficial veins* are immediately under the skin and are the cephalic, basilic and median veins and their tributaries.

The *deep veins* follow the course of the arteries and pour their blood into the *axillary vein*, which is a continuation of the basilic veins, and which becomes continuous with the subclavian veins.

The veins of the lower limbs are divided, like those of the upper limbs, into superficial and deep, which lie under the skin or accompany the arteries respectively.

The *superficial veins* are the short and long saphenous veins which empty into the deep popliteal vein at the back of the knee.

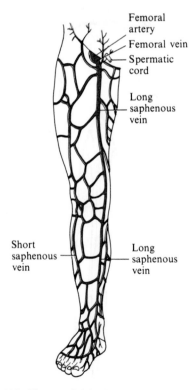

Femoral artery
Femoral vein
Spermatic cord
Long saphenous vein
Short saphenous vein
Long saphenous vein

FIG. 112. The superficial veins of the lower limb.

The *deep* veins are the anterior and posterior tibial veins, the popliteal vein and the femoral vein.

Between the deep and superficial veins of the lower limbs a number of 'perforating' veins exist which have their valves so arranged that normally blood is prevented from flowing from the

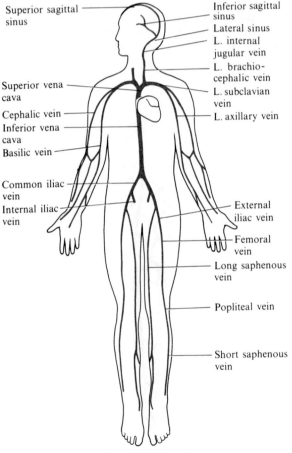

Superior sagittal sinus

Inferior sagittal sinus

Lateral sinus

L. internal jugular vein

L. brachio-cephalic vein

Superior vena cava

L. subclavian vein

Cephalic vein

L. axillary vein

Inferior vena cava

Basilic vein

Common iliac vein

Internal iliac vein

External iliac vein

Femoral vein

Long saphenous vein

Popliteal vein

Short saphenous vein

FIG. 113. Main veins.

deep to the superficial veins. If these valves become ineffective blood can flow from the deep to the superficial veins raising the pressure in the latter and causing the dilation and degeneration known as varicose veins and ulcers.

The femoral vein ends at the inguinal ligament by becoming the *external iliac vein*. The *internal iliac vein*, returning blood from the pelvis, unites with the external iliac vein to become the common iliac vein, one on either side of the body. These unite at the level of the fifth lumbar vertebra to form the *inferior vena cava* which receives several tributaries, some important ones being:

1. The *renal veins* from the kidneys.
2. The *ovarian* or *testicular* veins from the reproductive organs, which empty into the inferior vena cava on the right and into the renal vein on the left.
3. The *hepatic veins* which return all the blood from the liver, including that from the portal circulation.

The venae cavae pour all venous blood into the right atrium except that from the coronary circulation which is conveyed by the coronary sinus directly into the right atrium.

The Pulmonary Circulation (see also Chapter 16)

These vessels are concerned with carrying deoxygenated blood from the heart to the lungs and oxygenated blood from the lungs back to the heart. The *pulmonary trunk* carries venous blood from the right ventricle; at the level of the fifth thoracic vertebra it divides into *right* and *left pulmonary arteries* which then branch to carry blood to the various segments of the lungs. These vessels are the only arteries carrying deoxygenated blood.

The four *pulmonary veins* return the oxygenated blood from the lungs to the left atrium, two vessels from each lung. These are the only veins which carry oxygenated blood.

The Portal Circulation

The portal system includes all the veins which drain blood from the abdominal part of the digestive system and from the spleen, pancreas and gall bladder. The blood from these organs is carried to the liver by the *portal vein*, which ends in vessels like capillaries, called *sinusoids*. The blood is then conveyed by the hepatic veins to the inferior vena cava. As mentioned above the liver must also receive oxygenated blood and this is conveyed to it by the hepatic artery.

The Fetal Circulation

Fetal blood is carried to and from the placenta by umbilical arteries and veins. Most of the blood which enters the right atrium through the inferior vena cava passes through an opening in the atrial septum, called the *foramen ovale*, directly into the left atrium and thence into the left ventricle and aorta. The blood returning to the right atrium through the superior vena cava passes into the right ventricle and pulmonary trunk, after which only a small part of it goes to the lungs. Most of it passes through the *ductus arteriosus* directly to the aorta.

At birth the foramen ovale closes, so that blood cannot pass from the right atrium to the left atrium but is directed into the pulmonary trunk and thence to the lungs. The ductus arteriosus also closes shortly after birth.

15 The Lymphatic System

As blood passes through the capillaries in the tissues fluid oozes out through the porous walls and circulates through the tissues themselves, bathing every cell. This fluid is called *tissue* or *interstitial fluid*; it fills the interstices or the spaces between the cells which form the different tissues. It is clear, watery, straw-coloured fluid similar to the plasma of the blood from which it is derived. While blood circulates only through the blood vessels, tissue fluid circulates through the actual tissue and carries food, oxygen and water from the blood stream to each individual cell and carries away its waste products such as carbon dioxide, urea and water, transmitting them to the blood. It is, in other words, the carrying medium between the tissue cells and the blood.

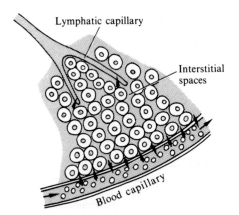

Fig. 114. Diagram showing the circulation of tissue fluid. Fluid passes out into the tissues and is collected partly by the blood capillary and partly by the lymph capillary.

Of the fluid which escapes from the capillaries into the tissues a certain amount passes back through the capillary wall, but its return is more difficult than its escape owing to the constant stream of oncoming blood which fills the capillaries. The excess fluid which cannot return directly into the blood stream is collected up and returned to the blood by a second set of vessels, which form the lymphatic system, and the fluid that these vessels contain is called *lymph*.

The lymphatic system consists of:

1. Lymphatic capillaries
2. Lymphatic vessels
3. Lymphatic nodes
4. Lymphatic ducts

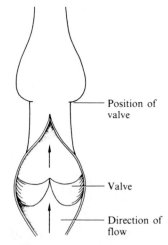

Position of valve

Valve

Direction of flow

Fig. 115.   Lymphatic vessel (cut open).

Lymphatic Capillaries

The lymphatic capillaries arise in the spaces in the tissues as fine hair-like vessels with porous walls. These gather up excess fluid from the tissues and unite to form the lymphatic vessels. The walls of lymph capillaries are permeable to substances of greater molecular size than those which can pass through the walls of blood capillaries.

Lymphatic Vessels

The lymphatic vessels are thin-walled, collapsible tubes similar in structure to the veins, but carrying lymph instead of blood. They are finer and more numerous than the veins and, like them are provided with valves to prevent the lymph moving in the wrong direction. Lymphatic vessels are found in most tissues except the

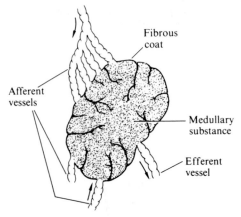

Fig. 116. A lymphatic node showing afferent and efferent vessels.

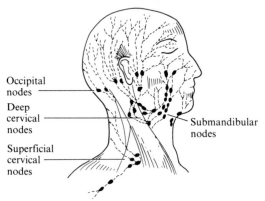

Fig. 117. The lymphatic vessels of the head and neck.

central nervous system, but run particularly in the subcutaneous tissues and pass through one or more lymphatic nodes.

Lymphatic Nodes

The lymphatic nodes are small bodies varying in size from a pin-head to an almond. Lymphatic vessels bring lymph to them and are called afferent vessels. These enter and divide up within the node and discharge the lymph into substance. The lymph is then gathered up again into fresh lymphatic vessels called efferent vessels, which carry it on and ultimately empty it into the lymphatic ducts after

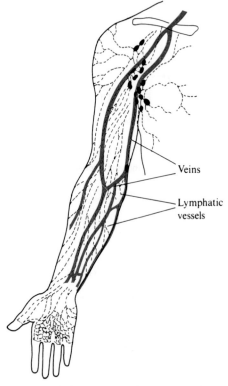

Veins

Lymphatic vessels

Fig. 118. The lymphatic vessels of the upper limb.

possibly carrying the lymph through more nodes. The lymphatic nodes consist essentially of cells similar to the white blood corpuscles (lymphocytes), held together by a network of connective tissue, which also forms a capsule for the node.

The *functions of the lymphatic nodes* are:

1. To *filter* the lymph of bacteria as it passes through. Thus when tissues are infected the node may become swollen and tender.

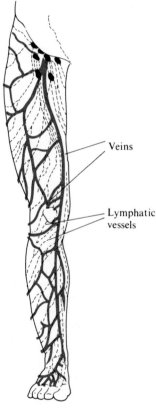

Veins

Lymphatic vessels

Fig. 119. The lymphatic vessels of the lower limb.

If the infection is a mild one, the organisms will be overcome by cells of the node and the tenderness and swelling will subside. If it is severe the organisms will cause acute inflammation and destruction of the white cells may result, causing an abscess to form in the node. If the bacteria are not destroyed by the node they may pass on in the lymph stream and infect the general circulation, causing septicaemia.

2. To provide fresh lymphocytes for the blood stream. The cells of the node constantly multiply and the newly formed cells are carried away in the lymph.

3. They may produce some antibodies and antitoxins to prevent infection.

The lymphatic nodes are for the most part massed together in groups in various parts of the body. Groups in the neck and under the chin filter the lymph from the head, tongue and the floor of the mouth. A group in the axilla filters the lymph from the upper limb and the chest wall. A group in the groin filters lymph from the lower limb and lower abdominal wall. Groups within the thorax and abdomen filter the lymph from the internal organs.

Special areas where much lymphatic tissue is found include: the palatine and pharyngeal tonsils; the thymus gland; aggregated lymphatic follicles in the small intestine; the appendix and the spleen.

The Lymphatic Ducts

After filtration by the nodes the lymph is emptied by the lymphatic vessels into the two lymphatic ducts, the thoracic duct and right lymphatic duct.

The *thoracic duct* is the larger. It begins in a small pouch at the back of the abdomen called the *cisterna chyli*. Into this empty all the lymphatic vessels from the lower limbs and the abdominal and pelvic organs. From the cisterna chyli the duct runs up through the mediastinum behind the heart to the root of the neck, where it turns to the left, is joined by the lymphatic vessels from the left side of the head and thorax and the left upper limb, and finally empties into the *left subclavian vein* at its junction with the left internal jugular vein. It is about 45 cm long and is provided with valves to prevent the lymph flowing in the wrong direction.

The *right lymphatic duct* is a comparatively small vessel formed by the joining of the lymphatic vessels from the right side of the

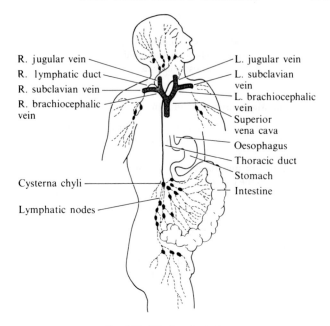

R. jugular vein
R. lymphatic duct
R. subclavian vein
R. brachiocephalic vein

L. jugular vein
L. subclavian vein
L. brachiocephalic vein
Superior vena cava
Oesophagus
Thoracic duct
Stomach
Intestine

Cysterna chyli

Lymphatic nodes

Fig. 120. The lymphatic ducts.

head and thorax and the right upper limb at the root of the neck. It is only about 1 cm long and enters into the *right subclavian vein*, where it joins the right internal jugular vein.

The lymphatic ducts thus gather up all the lymph and return it to the blood stream, from which the fluid in the tissues is constantly renewed.

The Functions of the Lymphatic System

The functions of the lymphatic system are as follows:

1. The lymphatic vessels gather up the excessive fluid or lymph from the tissues and thus permit a constant stream of fresh fluid to circulate through them.

2. It is the channel by which excess proteins in the tissue fluid pass back into the blood stream.

3. The nodes filter the lymph of bacterial infection and harmful substances.

4. The nodes produce fresh lymphocytes for the circulation.

5. The lymphatic vessels in the abdominal organs assist in the absorption of digested food, especially fat.

The Mechanism of the Lymphatic Circulation

The lymphatic circulation is maintained partly by suction, partly by pressure. Suction is the more important factor. The lymphatics empty into the large veins approaching the heart, and here there is negative pressure due to suction as the heart expands, and also suction towards the thorax during the act of inspiration.

Pressure is exerted on the lymphatics, as on the veins, by the contraction of muscles, and this outside pressure drives lymph onwards because the valves prevent a backward flow. There is also a slight pressure from the fluid in the tissues, on account of the constant pouring out of fresh fluid from the capillaries. If there is obstruction to the flow of the lymph through the lymphatic system, oedema results, i.e. a swelling of the tissues due to excess fluid collecting in them. The condition also results from any obstruction of the veins, since these also drain fluid from the tissues.

The Spleen

The spleen is a large nodule of lymphoid tissue. By function it belongs to the circulatory system, as do the lymphatic nodes. It is a deep purplish-red colour, and lies high up at the back of the abdomen on the left side behind the stomach. It is enclosed in a capsule of fibrous tissue, and fibrous strands make a supporting meshwork throughout the gland. The spaces of this meshwork are filled in with a pulp-like material, called the splenic pulp, which is the essential substance of the organ and contains cells of different types. Many of these are similar to the lymphocytes of the blood and lymph nodes, and these help to produce fresh white cells for the blood stream. Others are phagocytes or devouring cells which engulf the red blood cells that are beginning to wear out and break them down.

The *functions of the spleen* are not fully known, but it is thought to be:

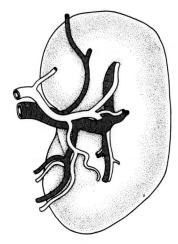

FIG. 121. The spleen and its blood vessels.

1. A source of fresh lymphocytes for the blood stream
2. A seat for the destruction of red blood cells

The spleen is also thought to assist in fighting infection, as it becomes enlarged in certain diseases where the blood is infected, e.g. malaria and typhoid fever. It probably helps in the manufacture of antibodies to fight infection. It is not essential to life, and can be removed by operation when it is responsible for ill health, e.g. in haemolytic anaemia.

16 The Respiratory System

All living cells require a constant supply of oxygen in order to carry on their metabolism. Oxygen is in the air, and the respiratory system is constructed in such a way that air can be taken into the lungs, where some of the oxygen is extracted for use by the body and at the same time carbon dioxide and water vapour are given up. The organs of the respiratory system are:

The nose
The pharynx
The larynx   } leading to the lungs
The trachea

The bronchi
The bronchioles   } within the lungs
The alveolar ducts and alveoli

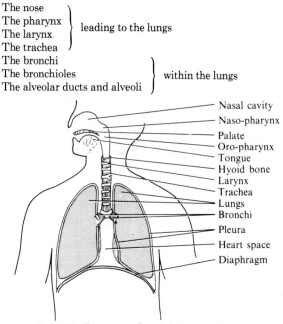

Nasal cavity
Naso-pharynx
Palate
Oro-pharynx
Tongue
Hyoid bone
Larynx
Trachea
Lungs
Bronchi
Pleura
Heart space
Diaphragm

Fig. 122. Diagram of the respiratory tract.

The Nose

The external nose is the visible part of the nose, formed by the two nasal bones and by cartilage. It is both covered and lined by skin and inside there are hairs which help to prevent foreign material from entering. The nasal cavity is a large cavity divided by a septum. The anterior nares are the openings which lead in from without and the posterior nares are similar openings at the back, leading in to the pharynx. The roof is formed by the ethmoid bone at the base of the skull and the floor by the hard and soft palates at the roof of the mouth. The lateral walls of the cavity are formed by the maxilla, the superior and middle nasal conchae of the ethmoid bone and the inferior nasal concha. The posterior part of the dividing septum is formed by the perpendicular plate of the ethmoid bone and by the vomer, while the anterior part is made of cartilage.

The three nasal conchae project into the nasal cavity on each side and greatly increase the surface area of the inside of the nose. The cavity of the nose is lined throughout with ciliated mucous membrane, which is extremely vascular, and atmospheric air is warmed as it passes over the epithelium, which contains many capillaries. The mucus moistens the air and entraps some of the dust and the cilia move the mucus back into the pharynx for swallowing or expectoration. The nerve endings of the sense of smell are situated in the highest part of the nasal cavity round the cribriform plate of the ethmoid bone.

Some of the bones surrounding the nasal cavity are hollow. The hollows in the bones are called the *paranasal sinuses*, which both lighten the bones and act as sounding chambers for the voice, making it resonant. The maxillary sinus lies below the orbit and opens through the lateral wall of the nose. The frontal sinus lies above the orbit towards the midline of the frontal bone. The ethmoidal sinuses are numerous and are contained within the part of the ethmoid bone separating the orbit from the nose, and the sphenoidal sinus is in the body of the sphenoid bone. All the paranasal sinuses are lined with mucous membrane and all open into the nasal cavity, from which they may become infected.

The Pharynx

The roof of the pharynx is formed by the body of the sphenoid bone and inferiorly it is continuous with the oesophagus. At the back it is separated from the cervical vertebrae by loose connective tissue,

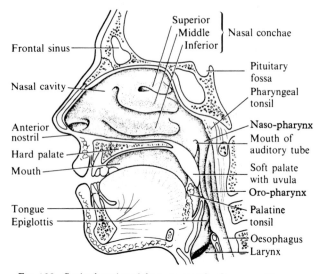

FIG. 123. Sagittal section of the nose, mouth, pharynx and larynx.

while the front wall is incomplete and communicates with the nose, mouth and larynx. The pharynx is divided into three sections, the naso-pharynx which lies behind the nose, the oro-pharynx which lies behind the mouth, and the laryngeal pharynx which lies behind the larynx.

The *naso-pharynx* is that part of the pharynx which lies behind the nose above the level of the soft palate. On the posterior wall there are patches of lymphoid tissue called the pharyngeal tonsils, commonly referred to as adenoids. This tissue sometimes enlarges and blocks the pharynx causing mouth breathing in children. The auditory tubes open from the lateral walls of the naso-pharynx and through them air is carried to the middle ear. The naso-pharynx is lined with ciliated mucous membrane which is continuous with the lining of the nose.

The *oro-pharynx* lies behind the mouth below the level of the soft palate, with which its lateral walls are continuous. Between the folds of these walls, which are called the palato-glossal arches, are collections of lymphoid tissue called the palatine tonsils. The oro-pharynx is part of both the respiratory tract and the alimentary tract, but it cannot be used for swallowing and breathing simultaneously. During the act of swallowing breathing stops momentarily and the oro-pharynx is completely blocked off from the naso-pharynx by the raising of the soft palate. The oro-pharynx is lined with stratified epithelium.

The Larynx

The larynx is continuous with the oro-pharynx above and with the trachea below. Above it lie the hyoid bone and the root of the tongue. The muscles of the neck lie in front of the larynx, and behind the larynx lies the laryngo-pharynx and the cervical vertebrae. On either side are the lobes of the thyroid gland. The larynx is composed of several irregular cartilages joined together by ligaments and membranes.

The *thyroid cartilage* is formed of two flat pieces of cartilage fused together in the front to form the laryngeal prominence or Adam's apple. Above the prominence is a notch called the thyroid notch. The thyroid cartilage is larger in the male than in the female.

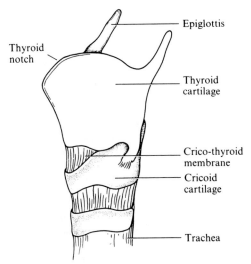

FIG. 124. Laryngeal cartilages.

The upper part is lined with stratified epithelium and the lower part with ciliated epithelium.

The *cricoid cartilage* lies below the thyroid cartilage and is shaped like a signet ring with the broad portion at the back. It forms the lateral and posterior walls of the larynx and is lined with ciliated epithelium.

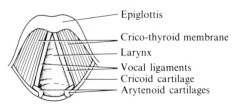

Epiglottis
Crico-thyroid membrane
Larynx
Vocal ligaments
Cricoid cartilage
Arytenoid cartilages

Vocal ligaments resting

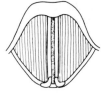

Vocal ligaments during speech

FIG. 125. Diagram of vocal ligaments.

The *epiglottis* is a leaf-shaped cartilage attached to the inside of the front wall of the thyroid cartilage immediately below the thyroid notch. During swallowing the larynx moves upwards and forwards so that its opening is occluded by the epiglottis.

The *arytenoid cartilages* are a pair of small pyramids made of hyaline cartilage. They are situated on top of the broad part of the cricoid cartilage, and the vocal ligaments are attached to them. They form the posterior wall of the larynx.

The hyoid bone and the laryngeal cartilages are joined together by ligaments and membranes. One of these, the *cricothyroid membrane*, is attached all round to the upper edge of the cricoid cartilage and has a free upper border, which is not circular like the lower border, but makes two parallel lines running from front to back. The two parallel edges are the *vocal ligaments*. They are fixed to the middle of the thyroid cartilage in front and to the arytenoid cartilages behind, and they contain much elastic tissue. When the intrinsic muscles of the larynx alter the position of the arytenoid cartilages the vocal ligaments are pulled together, narrowing the gap between them. If air is forced through the narrow gap, called the chink, during expiration, the vocal ligaments vibrate and sound is produced. The pitch of the sound produced depends on the length and tightness of the ligaments; an increased tension gives a higher note, a slacker tension a lower note. Loudness depends on the force with which the air is expired. The alteration of the sound into different words depends on the movements of the mouth, tongue, lips and facial muscles.

The Trachea

The trachea begins below the larynx and runs down the front of the neck into the chest. It divides into the right and the left main bronchi at the level of the fifth thoracic vertebra. It is about 12 cm long. The isthmus of the thyroid gland crosses in front of the upper part of the trachea, and the arch of the aorta lies in front of the lower part, with the manubrium of the sternum in front of it. The oesophagus lies behind the trachea, separating it from the bodies of the thoracic vertebrae. On either side of the trachea lie the lungs, with the lobes of the thyroid gland above them. The wall of the trachea is made of involuntary muscle and fibrous tissue strengthened by incomplete rings of hyaline cartilage. The deficiency in the cartilage lies at the back where the trachea is in contact with the oesophagus. When a bolus of food is swallowed the oesophagus is able to expand without hindrance, but the cartilage maintains the patency of the airway. The trachea is lined with ciliated epithelium containing goblet cells which secrete mucus. The cilia sweep the mucus and foreign particles upwards towards the larynx.

The Lungs

The lungs are two large spongy organs lying in the thorax on either side of the heart and great vessels. They extend from the root of the neck to the diaphragm and are roughly cone-shaped with the apex above and the base below. The ribs, costal cartilages and intercostal muscles lie in front of the lungs and behind them are the ribs, the intercostal muscles and the transverse processes of the thoracic vertebrae. Between the lungs is the *mediastinum*, a block of tissue which completely separates one side of the thoracic cavity from the other, stretching from the vertebrae behind to the sternum in front. Within the mediastinum lie the heart and great vessels, the trachea and the oesophagus, the thoracic duct and the thymus gland. The lungs are divided into lobes. The left lung has two lobes, separated by the oblique fissure. The superior lobe is above and in front of the inferior lobe which is conical in shape. The right lung has three lobes. The inferior lobe is separated by an oblique fissure in a similar manner to the left inferior lobe. The remainder of the lung is separated by a horizontal fissure into the superior lobe and the middle lobe. Each lobe is further divided into named broncho-pulmonary segments, separated from each other by a wall of connective tissue and each having an artery and a vein. Each segment is also divided into smaller units called lobules.

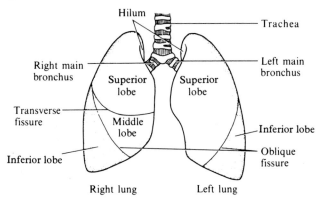

Fig. 126. Diagram showing lobes of lungs.

The Bronchi

The two main bronchi commence at the bifurcation of the trachea and one leads into each lung. The left main bronchus is narrower, longer and more horizontal than the right main bronchus because the heart lies a little to the left of the midline. Each main bronchus divides into branches, one for each lobe. Each of these then divides into named branches, one for each broncho-pulmonary segment, and divides again into progressively smaller bronchi within the lung substance. The bronchi are similar in structure to the trachea but the cartilage is less regular.

The Bronchioles. The finest bronchi are called bronchioles. They have no cartilage but are composed of muscular, fibrous and elastic tissue lined with cuboid epithelium. As the bronchioles become smaller, the muscular and fibrous tissue disappears and the smallest tubes, called terminal bronchioles, are a single layer of flattened epithelial cells.

The Alveolar Ducts and Alveoli. The terminal bronchioles branch repeatedly to form minute passages called alveolar ducts, from which alveolar sacs and alveoli open. The alveoli are surrounded by a network of capillaries. Deoxygenated blood enters the capillary network from the pulmonary artery and oxygenated blood leaves it to enter the pulmonary veins. It is in the capillary network that the exchange of gases takes place between the air in the alveoli and the blood in the vessels.

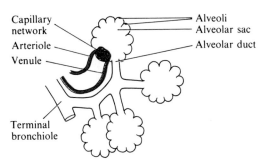

FIG. 127. Diagram of alveoli.

The Hilum of the Lung

The hilum is a triangular-shaped depression on the concave medial surface of the lung. The structures forming the root of the lung enter and leave at the hilum, which is at the level of the fifth to seventh thoracic vertebrae. These structures include the main bronchus, the pulmonary artery, the bronchial artery and branches of the vagus nerve, which enter at this point, and two pulmonary veins, the bronchial veins and lymphatic vessels, which leave the lung at the root. There are also many lymph nodes round the root of the lung.

The Pleura

The pleura is a serous membrane which surrounds each lung. It is composed of flattened epithelial cells on a basement membrane and it has two layers. The visceral pleura is firmly attached to the lungs, covering their surfaces and dipping into the inter-lobar fissures. At the root of the lungs the visceral layer is reflected back to become the parietal layer which lines the chest wall and covers the superior surface of the diaphragm. The two layers of the pleura are normally in close contact with each other, separated only by a film of serous fluid which enables them to glide over one another without friction. This potential space between the layers is called the pleural cavity.

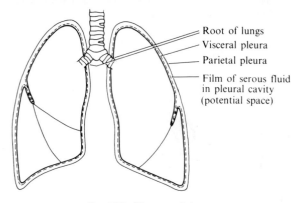

Root of lungs
Visceral pleura
Parietal pleura
Film of serous fluid in pleural cavity (potential space)

Fig. 128. Diagram of pleura.

Gaseous Exchange

Exchange of gases within the body takes place both in the lungs, where it is called external respiration, and in the tissues, where it is called internal respiration. An elementary law of physics relating to gases states that they tend to diffuse from a higher pressure to a lower pressure. The air which is breathed in contains several gases. The composition of inspired air is:

Nitrogen, 79 per cent
Oxygen, 21 per cent
Carbon dioxide, 0·04 per cent
Water vapour
Traces of other gases

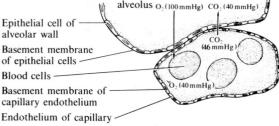

FIG. 129. Diagram of gaseous exchange in the lungs.

External Respiration. When inspired air reaches the alveoli it is in close contact with the blood in the surrounding capillary network. Oxygen in the alveoli, at a pressure of 100 mmHg, comes in contact with oxygen in the venous blood at a pressure of 40 mmHg and therefore the gas diffuses into the blood until the pressures are equal. At the same time carbon dioxide in the blood, at a pressure of 46 mmHg, comes in contact with alveolar carbon dioxide at a pressure of 40 mmHg and therefore the gas diffuses out of the blood into the alveoli. The gaseous content of expired air is therefore altered to contain less oxygen and more carbon dioxide. The nitrogen content remains the same. The composition of expired air is:

Nitrogen, 79 per cent
Oxygen, 16 per cent
Carbon dioxide, 4·5 per cent
Water vapour
Traces of other gases

Internal Respiration. The oxygen which has diffused into the blood is carried in the haemoglobin—now called oxyhaemoglobin—to the tissues. Here the pressure of the oxygen is low, so the gas diffuses out of the blood and into the tissues, the amount depending on the activity in the tissue. At the same time carbon dioxide, produced in the tissues, is carried away by the blood.

Mechanism of Respiration

Respiration consists of two parts, inspiration and expiration. The chest expands during inspiration owing to movement of the diaphragm and the intercostal muscles. When the diaphragm contracts during inspiration it is flattened and lowered and the thoracic cavity is increased in length (see Chapter 11). The external intercostal muscles, on contraction, lift the ribs and draw them out, increasing the depth of the thoracic cavity. As the chest wall moves up and out the parietal pleura, which is closely attached to it, moves with it. The visceral pleura follows the parietal pleura and the volume of the interior of the thorax is increased. The lung expands to fill the space and air is sucked into the bronchial tree.

Expiration during quiet breathing is passive. The diaphragm relaxes and assumes its original domed shape. The intercostal muscles relax and the ribs revert to their previous position. The lungs recoil and air is driven out through the bronchial tree. In forced expiration the internal intercostal muscles contract actively to lower the ribs. The accessory muscles of respiration may be brought into use during deep breathing or when the airway is obstructed. During inspiration the sternocleidomastoid muscles raise the sternum and increase the diameter of the thorax from front to back. Serratus anterior and pectoralis major pull the ribs outwards when the arm is fixed. Latissimus dorsi and the muscles of the anterior abdominal wall help to compress the thorax during forced expiration.

Control of Respiration

Respiration is controlled by the respiratory centre in the medulla oblongata. The presence of an accumulation of carbon dioxide in the blood stimulates specialized cells in the great arteries. Impulses are carried by the vagus and glossopharyngeal nerves to the respiratory centre, from the respiratory centre by the phrenic nerves to the diaphragm and by the intercostal nerves to the intercostal muscles. These impulses cause the muscles to contract and inspiration occurs.

The Capacity of the Lungs

Tidal volume is the amount of air breathed in and out during normal quiet breathing (about 500 ml). After a normal expiration the amount of air taken in during forced inspiration is called the *inspiratory capacity* (about 3000 ml). It includes the tidal volume. After a quiet expiration it is possible to force another 1000 ml of air from the lungs. This is the *expiratory reserve volume.*

Table 12
Lung Capacity

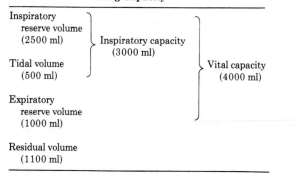

Inspiratory reserve volume (2500 ml)	Inspiratory capacity (3000 ml)	Vital capacity (4000 ml)
Tidal volume (500 ml)		
Expiratory reserve volume (1000 ml)		
Residual volume (1100 ml)		

After the deepest possible expiration there is still the *residual volume* (about 1100 ml) left in the respiratory passages.

Vital capacity is the largest volume of air expired after the deepest possible inspiration (about 4000 ml). An important measure of respiratory function is the amount of forced expiration in one second. It is called the *forced expiratory volume* and should measure 75 to 80 per cent of the vital capacity.

17 The Digestive System

The digestive system consists of all the organs which are concerned in the chewing, swallowing, digestion and absorption of food and in the elimination from the body of indigestible and undigested food.

It consists of the digestive tube or alimentary canal and the accessory organs of digestion.

The *digestive tube* is about nine metres long and consists of the following parts:

1. The mouth
2. The pharynx
3. The oesophagus
4. The stomach
5. The small intestine
6. The large intestine, which reaches the surface of the body at the anus

The *accessory organs* are:

1. The teeth
2. Three pairs of salivary glands
3. The liver and bile ducts (see Chapter 18)
4. The pancreas (see Chapter 18)

The Mouth

The mouth is a cavity bounded externally by the lips and cheeks and leading into the pharynx. The roof is formed by the hard and soft palates and the anterior two-thirds of the tongue fills the floor of the mouth. The walls are formed by the muscles of the cheeks. The mucous membrane which lines the mouth is continuous with the skin of the lips and with the mucous lining of the pharynx. The lips enclose the orbicularis oris muscle which keeps the mouth closed.

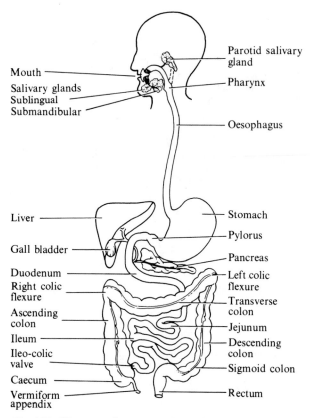

Fig. 130. Diagrammatic representation of the digestive system.

The *hard palate* is formed by parts of the palatine bones and the maxillae; its upper surface forms the floor of the nasal cavity. The *soft palate* is suspended from the posterior border of the hard palate and extends down between the oral and nasal parts of the pharynx. Its lower border hangs like a curtain between the mouth and the pharynx and a small conical process, called the *uvula*, hangs down from it. Two curved folds of mucous membrane extend sideways and downwards from each side of the base of the uvula, called the

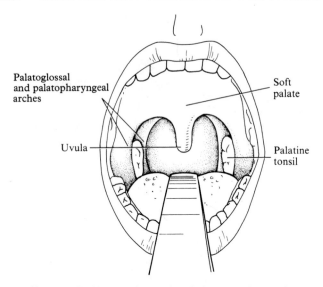

Palatoglossal and palatopharyngeal arches

Soft palate

Uvula

Palatine tonsil

FIG. 131. Looking into the mouth with the tongue depressed.

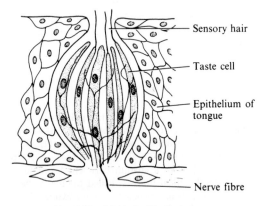

Sensory hair

Taste cell

Epithelium of tongue

Nerve fibre

FIG. 132. A taste bud in the tongue.

palatoglossal and *palatopharyngeal arches*, between which lie the masses of lymphoid tissue known as the *palatine tonsils*.

The Tongue

The tongue is a muscular organ which is attached to the hyoid bone and the mandible. It is covered in certain areas with modifications of the mucous membrane which appear as projections to increase the surface area and are called *papillae*. In addition specialized areas called *taste buds* are widespread over almost the entire area of the tongue. The under surface of the anterior part of the tongue is connected to the floor of the mouth by a fold of mucous membrane called the *frenulum*. The functions of the tongue are:

1. It is the organ of *taste*
2. It assists in the *mastication* of food
3. It assists in *swallowing*
4. It assists with *speech*

The Teeth

Man is provided with two sets of *teeth* which make their appearance at different periods of life. The first set are *deciduous* or *primary* teeth and erupt through the gums during the first and second years of life. The second set begin to replace the first about the sixth year and the process is usually complete by the twenty-fifth year. Since they cannot be replaced, and may be retained until old age, they are known as the *permanent* teeth.

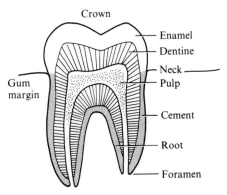

FIG. 133. Section of a tooth, showing its structure.

Each tooth consists of three parts:

1. The *crown*, projecting beyond the gum
2. The *root*, embedded in the alveolus of the maxilla or mandible—a tooth may have one, two or three roots
3. The *neck*, the constricted part between the crown and the root

In the centre of all these parts is the *pulp* and immediately outside the pulp is a yellowish-white layer, called *dentine*, which forms the main part of the tooth. The outer layer of the tooth is in two parts; that covering the crown is called *enamel* and is a hard, white layer while that covering the root is called *cement* and is a thin layer resembling bone in structure. The pulp is richly supplied with blood vessels and nerves which enter the tooth through foramina at the apex of each root.

There are four types of teeth:

1. The *incisor teeth* have chisel-shaped crowns giving a sharp cutting edge for biting food.
2. The *canine teeth* have large conical crowns.
3. The *premolar (or bicuspid) teeth* have almost circular crowns with two cusps for grinding food.
4. The *molar teeth* are the largest, and they have broad crowns with four or five cusps.

There are 20 deciduous teeth and 32 permanent teeth.

Table 13
The Deciduous Teeth

	Molars	Canines	Incisors		Canines	Molars
Upper jaw	2	1	2	2	1	2
Lower jaw	2	1	2	2	1	2

The Permanent Teeth

	Molars	Premolars	Canines	Incisors		Canines	Premolars	Molars
Upper jaw	3	2	1	2	2	1	2	3
Lower jaw	3	2	1	2	2	1	2	3

The lower central incisors are the first of the deciduous teeth to be cut at the age of about 6 to 8 months, though occasionally they are present at birth. The upper and lateral incisors follow by the age of about one year and all the teeth have generally erupted by the age of 2 to $2\frac{1}{2}$ years, though there is a wide variation in the ages at which teeth erupt.

By the sixth year the first molars have usually erupted and there are 24 teeth altogether. Next the deciduous incisors drop out and the permanent teeth, which have developed in the jaw throughout childhood, gradually replace them. The permanent teeth have usually erupted by the age of 14 years, with the exception of the third molar, or wisdom tooth, which does not erupt until between 18 and 25 years of age. Normally the grinding teeth in the upper and lower jaws do not lie exactly opposite each other because the upper incisors are wider than the lower. This enables the cusps of the upper teeth to grind effectively with the cusps of the corresponding teeth which are distal to them. The lower teeth are usually half a tooth's width in front of the upper.

For the development of good teeth it is important that the mother during pregnancy, and the child throughout the growing period, should be liberally supplied with foods rich in calcium, particularly milk and eggs, and with vitamin D, which controls bone formation. It is also important that the jaws are well exercised by breast feeding, followed by the giving of hard foods, so that the jaws and teeth receive a good blood supply. Poor enamel, due to lack of vitamin D or calcium, will result in early decay, as will also the collecting of starchy and sugary foods about the teeth, especially during the night. These foods, as they decompose in the mouth, produce acids which will act on the calcium of the teeth and make it soluble, causing the teeth to become softer and thus allowing bacteria to enter. Poor jaw development from lack of exercise will cause the teeth to be crowded and out of position, often overlapping; also there may not be room for the third molar tooth (the wisdom tooth) to come through and it may become impacted.

The Salivary Glands

There are three pairs of *salivary glands*. The *parotid gland* is the largest and lies just below the ear; its duct is about 5 cm long and opens into the mouth opposite the second upper molar tooth. It is this gland which is affected by the disease commonly known as mumps. The *submandibular gland* and the *sublingual gland* both open into the floor of the mouth. *Saliva* is secreted reflexly by the presence of food in the mouth and by a conditioned (or learned) reflex which enables saliva to be secreted by the sight, smell or thought of food. Saliva contains a large amount of water which moistens and softens the food; mucus, which combines the food and lubricates it for its passage down the oesophagus; and the enzyme *ptyalin*, which acts on cooked starch (carbohydrate) and splits it into maltose and dextrin. Saliva also cleanses the mouth and teeth and keeps the soft parts supple.

The Pharynx

When the food is well chewed and moistened the tongue rolls it into a *bolus* and carries it towards the *oral part of the pharynx*. The soft palate rises up to occlude the naso-pharynx and the epiglottis moves upwards and forwards so that the bolus passes over the closed inlet of the larynx and on into the *laryngeal part of the pharynx* and thence to the oesophagus. This is an excellent example of muscular coordination and if it is not achieved correctly 'choking' will result (see Fig. 123).

The Oesophagus

The oesophagus is a muscular canal about 25 cm long extending from the pharynx to the stomach. It begins at the level of the sixth cervical vertebra and descends through the mediastinum in front of

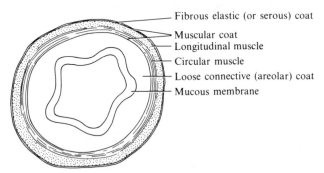

Fibrous elastic (or serous) coat
Muscular coat
Longitudinal muscle
Circular muscle
Loose connective (areolar) coat
Mucous membrane

FIG. 134.   Cross-section of the alimentary tract.

the vertebral column and behind the trachea. It passes through the diaphragm at the level of the tenth thoracic vertebra and ends at the cardiac orifice of the stomach at the level of the eleventh thoracic vertebra. On each side of the upper part of the oesophagus are the corresponding common carotid artery and part of the thyroid gland. The oesophagus has four coats and is similar in structure to the remainder of the alimentary canal.

1. The *fibrous outer coat* consists of areolar tissue containing many elastic fibres.

2. The *muscular coat* has two layers, the outer fibres running longitudinally and the inner layer consisting of circular fibres.

3. The *areolar* or *submucous coat* connects the mucous and muscular coats and contains the larger blood vessels and nerves, as well as the mucous glands.

4. The inner lining of *mucous membrane* secretes mucus.

The muscular coat of the upper two-thirds of the oesophagus is of striped voluntary muscle; the lower one-third contains unstriped involuntary muscle. The oesophagus is innervated by the vagus nerve. Movement of food through the oesophagus is by peristaltic action. *Peristalsis* means a wave of dilatation followed by a wave of contraction as succeeding muscle fibres relax and contract. It takes about nine seconds for a wave of peristalsis to pass the bolus of food from the pharynx to the stomach.

The Stomach

The stomach is the most dilated part of the digestive tube and is situated between the end of the oesophagus and the beginning of the small intestine. Its shape and position are altered by changes within the abdominal cavity and by the stomach contents but it lies below the diaphragm, slightly to the left of the midline.

The stomach is approximately J-shaped and has two curvatures. The *lesser curvature* forms the right (or posterior) border of the stomach. The *greater curvature* is directed mainly forwards and first forms an arch upwards and to the left to form the *fundus* of the stomach; it then passes downwards and finally turns right to the point where it joins the duodenum. The capacity of the stomach is about 1500 ml in the adult.

The upper opening from the oesophagus is called the *cardiac*

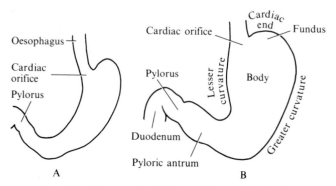

Fig. 135. The stomach. (A) Empty, (B) containing food.

orifice and the circular muscle fibres of the oesophagus are slightly thicker at this point and constitute a weak sphincter muscle.

The lower opening, into the duodenum, is called the *pyloric orifice* and it is guarded by the strong *pyloric sphincter* which prevents regurgitation of food from the duodenum into the stomach.

The wall of the stomach consists of four coats:

1. The outer *serous coat*, the visceral layer of the peritoneum (see page 207).

2. The *muscular coat* which consists of three layers of unstriped muscle fibres, the outer being longitudinal, the middle being circular and the inner being oblique.

3. The *submucous coat* of loose areolar tissue.

4. The lining of *mucous membrane*, which is honeycombed in appearance because of the presence of the gastric glands and their openings. The mucous membrane has numerous folds, called *rugae*, which run longitudinally and which flatten out when the stomach is full. The mucus secreted by the goblet cells helps to lubricate the food.

Functions of the Stomach

1. To churn up the food, breaking it up still further and mixing it with the secretions from the gastric glands. The part of the stomach just before the pyloric sphincter—the *pyloric antrum*—plays a large part in this movement, muscular contraction and relaxation sending some of the now liquid food through the sphincter into the small

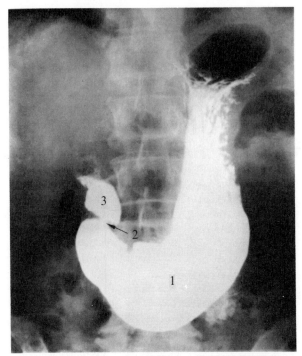

FIG. 136. The stomach. An X-ray of a barium meal, showing the normal stomach, the constriction of the pylorus and the 'duodenal cap', i.e. the beginning of the duodenum, filled with barium. (1) Body of stomach; (2) pylorus; (3) duodenal cap.

intestine and returning some to the body of the stomach for further mixing.

2. To continue the digestion of food by means of the gastric juice.

3. To secrete the intrinsic factor.

There are three types of cell in the mucosa of the stomach. *Mucous cells* secrete mucus which protects the mucous membrane from the action of the other gastric juices; *chief cells* secrete an enzyme known as pepsinogen, and in children another called rennin, and *oxyntic cells* secrete hydrochloric acid. The secretion of the

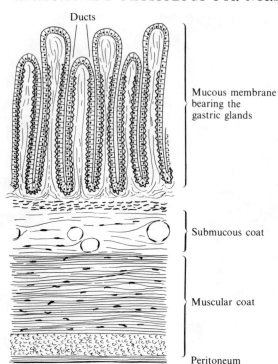

Ducts

Mucous membrane
bearing the
gastric glands

Submucous coat

Muscular coat

Peritoneum

Fig. 137. Section of gastric wall highly magnified to show glands.

gastric juice occurs reflexly in the same way as the saliva, causing a copious flow of the fluid before and during the taking of food. The gastric glands are also stimulated by an internal secretion or hormone, produced by the stomach, called *gastrin*, which passes into the circulation and when it reaches the gastric glands increases the production of gastric juice.

Gastric Juice. Gastric juice consists of:

1. Water, mineral salts and mucus.
2. Hydrochloric acid (HCl).
3. Pepsinogen, which is converted by hydrochloric acid into

the active enzyme pepsin. *Pepsin* turns proteins into peptones.

4. Rennin, which coagulates milk protein into casein, in which form it can be acted upon by pepsin.

The juice makes the food more liquid and acid in reaction. Until it becomes acid, and this may take 15 to 30 minutes in the cardiac end of the stomach, which acts as a reservoir, the *ptyalin* of the saliva continues to act on the cooked starch. When the food is acid, *pepsin* and *rennin* act on the proteins and on caseinogen. The food is quickly acidified in the pyloric end of the stomach, where peristaltic action is very marked, so that it acts like a mill, churning the food up and mixing it with the gastric juice. The food remains in the stomach for $\frac{1}{2}$ hour to 3 hours or more, according to the nature of the food and the muscularity of the individual stomach. A meal rich in carbohydrate but containing little protein, such as tea, toast and cake, will leave the stomach in half an hour. A good mixed meal, like a typical dinner, will remain for $2\frac{1}{2}$ to 3 hours or more, though it may leave earlier or stay longer according to the tone and activity of the muscular coat.

The hydrochloric acid in the gastric juice serves several purposes:

1. It gives the acid reaction required by the gastric enzymes
2. It kills bacteria
3. It controls the pylorus
4. It stops the action of ptyalin
5. It converts pepsinogen to pepsin

The pylorus is normally contracted. When there is food in the stomach the gastric juice makes the contents gradually more and more acid at the pyloric end. When it reaches a certain degree of acidity the pylorus relaxes and a little food passes into the duodenum. The acid food here causes the pylorus to close and the tone of the stomach wall drives food from the cardiac reservoir down to mix with the food in the pyloric end, making it less acid. Gradually the food in the duodenum is made alkaline and that in the pyloric end of the stomach again becomes more acid; this causes the pylorus to open again.

The churning action of the stomach serves to emulsify coarsely any fat which may be present and which the body heat will have melted. This converts the food into a greyish-white fluid called *chyme*.

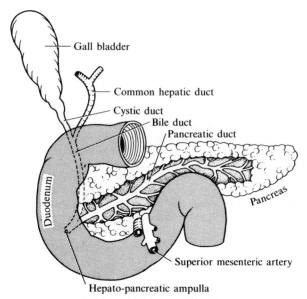

Gall bladder

Common hepatic duct

Cystic duct

Bile duct

Pancreatic duct

Duodenum

Pancreas

Superior mesenteric artery

Hepato-pancreatic ampulla

Fig. 138. The duodenum, pancreas and gall bladder.

The Small Intestine

The small intestine is a convoluted tube extending from the pyloric sphincter to its junction with the large intestine at the ileo-caecal valve. It is about 6 metres in length and lies in the central and lower parts of the abdominal cavity, usually within the curves of the large intestine (see Fig. 130). The small intestine consists of the duodenum, the jejunum and the ileum.

The *duodenum* is a short curved portion of about 25 cm long and is the widest and most fixed part of the small intestine. It is roughly C-shaped and curves round the head of the pancreas. Ducts from the gall bladder and liver and the pancreas enter the medial aspect of the duodenum through the *hepato-pancreatic ampulla*, which is guarded by a sphincter-like muscle.

The *jejunum* is the name given to the upper two-fifths of the remainder of the small intestine and the lower three-fifths is called the

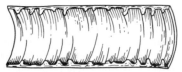

FIG. 139. Section of small intestine, showing puckered lining (circular folds).

ileum. Both are attached to the posterior abdominal wall by a fold of peritoneum called the mesentery (see Fig. 146).

The wall of the small intestine has the same four coats as the remainder of the alimentary tract.

1. Serous coat, formed of peritoneum
2. Muscular coat, with a thin external layer of longitudinal fibres and a thick internal layer of circular fibres
3. Submucous coat, containing blood vessels, lymph vessels and nerves
4. Mucous membrane lining

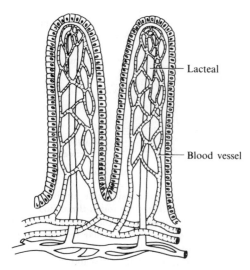

FIG. 140. Villi containing blood vessels and lacteal.

The mucous membrane lining has three special features:

1. It is thrown into circular folds which, unlike the rugae of the stomach, are permanent and are not obliterated when the intestine is distended. They increase the area available for absorption.

2. It has a velvety appearance due to the presence of fine hair-like projections called *villi*, each containing a lymph vessel called a *lacteal*, and blood vessels.

3. It is supplied with *glands* of the simple, tubular type which secrete *intestinal juice*.

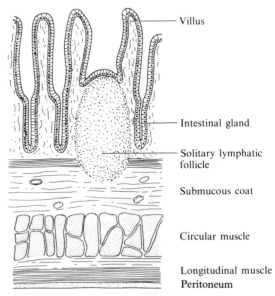

FIG. 141. Section of the wall of the small intestine, highly magnified.

The small intestine contains considerable amounts of lymphoid tissue. *Solitary lymphatic follicles* are found throughout the mucous membrane but are most numerous in the lower part of the ileum. *Aggregated lymphatic follicles* form circular or oval patches large enough to be seen with the naked eye. These nodules deal with bacteria which may be absorbed from the intestine; they become inflamed as a complication of such diseases as typhoid fever.

Functions of the Small Intestine

The functions of the small intestine are the digestion and absorption of food.

Digestion. Digestion is carried out by the pancreatic juice, bile and intestinal juice.

Table 14

Digestive Processes

Source	Secretion	Changes
Saliva	Ptyalin	Cooked starch into maltose and dextrin
Gastric juice	Pepsin	Proteins into peptones
	Rennin	Coagulates milk
	Hydrochloric acid	Stops action of ptyalin
		Converts pepsinogen to pepsin
Pancreatic juice	Trypsin	Proteins to polypeptides
	Amylase	All starch to maltose and dextrin
	Lipase	Fats to fatty acids and glycerol
Liver	Bile	Emulsifies fats
Intestinal juice	Enterokinase	Pancreatic trypsinogen to active trypsin
	Peptidase	Polypeptides to amino acids
	Maltase	Maltose ⎫
	Sucrase	Sucrose ⎬ to glucose
	Lactase	Lactose ⎭
	Lipase	Fats to fatty acids and glycerol

The pancreatic and intestinal juices are secreted reflexly through sensations concerning food, and also through a hormone called *secretin* produced by the lining of the intestine, as gastrin is secreted by the stomach wall. The juices are alkaline and make the food alkaline in reaction. This finely emulsifies the fat.

The pancreatic juice consists of water, alkaline salts and three enzymes acting on three different foodstuffs:

1. *Trypsinogen*, which turns peptones and proteins into amino acids when converted into active trypsin by enterokinase. If active trypsin were secreted by the pancreas, it would be able to digest the protein of the cells which form the gland and its ducts. Trypsin becomes active only when it mixes with the food and the intestinal juice within the bowel.

2. *Amylase*, which turns starch, cooked and uncooked, into malt sugar (maltose).

3. *Lipase*, which splits fat into fatty acids and glycerol, after the bile has emulsified the fat, to increase the surface area.

The *bile* contains no enzymes, but is rich in alkaline salts, which serve to emulsify and saponify fats, i.e. make soaps from them.

The *intestinal juice* contains water, salts and enzymes. These enzymes are:

1. *Enterokinase*, which converts trypsinogen secreted by the pancreas to active trypsin.

2. *Peptidase*, which acts on peptones and turns them into amino acids.

3. *Maltase*, which turns maltose into simple sugar, such as glucose.

4. *Sucrase*, which turns cane sugar (sucrose) into simple sugar.

5. *Lactase*, which turns lactose into simple sugar.

6. *Lipase*, which completes the conversion of fats to fatty acids and glycerol.

These juices are mixed with the food by *peristalsis*, the muscular action of the wall of the small intestine. These movements may be seen on a film taken with the small intestine exposed by an abdominal incision. The contractions occur first at one place and then at another and are followed by relaxation, having a kneading or churning effect and bringing the mucous lining into close contact with the contents of the gut.

Absorption. The absorption of proteins, carbohydrates and fats takes place almost entirely through the villi in the small intestine. Very little food is absorbed from the stomach, as it is not yet sufficiently digested, or, if absorbable, e.g. glucose and water, does not stay in the stomach, but merely passes through it. *Proteins* in the form of amino acids and *carbohydrates* in the form of simple sugar are absorbed by the cells covering the villi and pass into the *blood capillaries*, being carried by the portal vein to the liver. *Fats* in the form of fatty acids and glycerol are absorbed by the cells covering the villi and built up again by them into droplets of fat. These pass into the lymph within the villi and are drained away by the *lymphatic capillaries* or lacteals, which are so named because the lymph they contain is made milk-like by the droplets of fat in suspension. The fat passes via the lymphatic vessels to the cisterna chyli and is carried up the thoracic duct into the blood stream.

The Large Intestine

The large intestine extends from the end of the ileum to the anus and is about 1·5 metres long. It forms an arch which encloses most of the small intestine, and is divided into seven sections:

1. The caecum
2. The ascending colon
3. The transverse colon
4. The descending colon
5. The sigmoid colon
6. The rectum
7. The anal canal

The *caecum* lies in the right iliac fossa. It is a dilated area which has a blind lower end but is continuous above with the ascending colon and where one passes into the other the ileum opens into the caecum through the *ileo-caecal valve*. This valve is a sphincter and prevents the caecal contents passing back into the ileum. Taking food into the stomach initiates contraction of the duodenum and the rest of the small intestine, followed by the passage of the contents of the ileum into the caecum through the ileo-caecal valve. This is called the *gastro-ileal reflex*.

The *vermiform appendix* is a narrow blind-ended tube which opens out of the caecum about 2 cm below the ileo-caecal valve. It is usually about 9 cm long, though it can vary from 2 to 20 cm in length and can occupy a variety of positions within the abdomen. The submucous coat of the appendix contains considerable amounts of lymphoid tissue.

The *ascending colon* is about 15 cm in length and is narrower than the caecum. It ascends the right side of the abdomen to the under surface of the liver where it bends forwards and to the left at the *right colic flexure*.

The *transverse colon* is about 50 cm in length and passes across the abdomen to the under surface of the spleen in an inverted arch. Here it curves sharply downwards at the *left colic flexure*.

The *descending colon* is about 25 cm in length and passes down the left side of the abdomen to the inlet of the lesser pelvis, where it becomes the *sigmoid colon*.

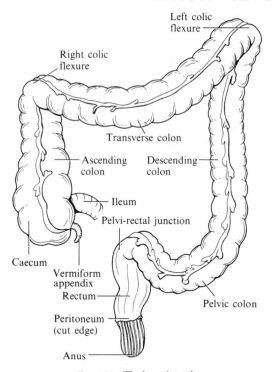

Fig. 142. The large intestine.

The *sigmoid colon* forms a loop which is about 40 cm in length and lies within the lesser pelvis.

The *rectum* is continuous above with the sigmoid colon. It is about 12 cm long and passes through the pelvic diaphragm to become the anal canal.

The *anal canal* passes downwards and backwards to end at the anus. At the junction of the anus and the rectum the unstriped circular muscle becomes thickened to form the *internal anal sphincter* which surrounds the upper three-quarters of the anal canal. The *external anal sphincter* surrounds the whole length of the anal

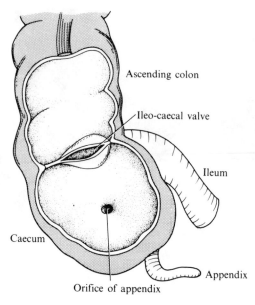

FIG. 143. The caecum, with the appendix.

canal and it is the tone of these sphincters which keep the anal canal and the anus closed. The external sphincter can be contracted voluntarily to close the anus more firmly.

The wall of the large intestine has the same four coats as the remainder of the alimentary tract:

1. The outer serous coat of peritoneum.
2. The muscular coat of external longitudinal and internal circular fibres. The longitudinal fibres form a continuous layer, but in some places the layer is thickened to form three bands called *taeniae coli*. The bands are shorter than the other coats of the large intestine and produce the typical puckered or sacculated appearance. The sacculations are called the *haustrations*.
3. The submucous coat.
4. The lining of mucous membrane.

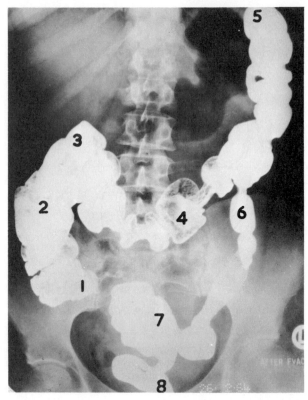

Fig. 144. The colon. (1) Caecum, with appendix; (2) ascending colon; (3) right colic flexure; (4) transverse colon; (5) left colic flexure; (6) descending colon; (7) sigmoid flexure; (8) beginning of the rectum.

Functions of the Large Intestine

    1. To absorb water and salts
    2. To excrete faeces

The material which enters the large intestine consists of water, salts, very little food material, as this has been digested and absorbed in the small intestine, cellulose, which is indigestible, and bacteria. The bacteria are very numerous; although they are largely

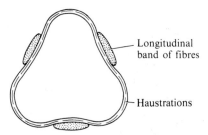

FIG. 145. Cross-section of the large intestine.

killed in the stomach, the alkaline reaction, food, warmth and moisture in the small intestine encourage their growth here. The material is in a very fluid state. In the colon water and salts are quickly absorbed, so that the fluid is rapidly turned into a paste containing the cellulose and bacteria, many of which die from lack of water and food. This paste forms the faeces, which consists of a fluid and solid part, the latter being roughly 50 per cent cellulose and 50 per cent dead bacteria. At intervals mass movements propel the faeces into the rectum, from which it is excreted.

Movements of the colon are similar to those seen in the small intestine but peristalsis occurs less frequently. A very strong wave of peristalsis occurs three to four times a day and moves material on into the pelvic colon.

Defaecation, or the passage of faeces, occurs because of the movement of food residue from the pelvic colon into the rectum, which becomes distended. This distension causes a reflex contraction of the rectal muscles which tends to expel the contents through the anus, though this depends on the relaxation of the external anal sphincter. Defaecation is therefore a reflex action which can be inhibited voluntarily. The gastro-ileal reflex also causes emptying of the rectum (see page 203). If defaecation is delayed the sensation of fullness in the rectum passes off, more water is absorbed from the faeces through the rectal wall and constipation may occur.

The Peritoneum

The peritoneum is a serous membrane and in the male is a closed sac lining the abdomen. In the female the free ends of the uterine tubes open into the peritoneal cavity. The part which lines the

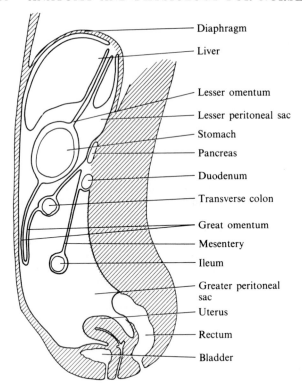

Diaphragm

Liver

Lesser omentum

Lesser peritoneal sac

Stomach

Pancreas

Duodenum

Transverse colon

Great omentum

Mesentery

Ileum

Greater peritoneal sac

Uterus

Rectum

Bladder

FIG. 146. The peritoneum.

abdominal wall is named the *parietal portion* of the peritoneum; that part which is reflected over the organs is called the *visceral portion*. The two layers are in contact with each other but the potential space between them is named the *peritoneal cavity*. This cavity consists of the greater peritoneal sac and the lesser peritoneal sac.

Other specially named areas are the *greater omentum*, a double fold of peritoneum which descends from the lower border of the stomach and loops up again to the transverse colon. This helps to prevent the spread of infection from the organs into the peritoneum. The *lesser omentum* is a fold which extends to the liver from the lesser curvature of the stomach and the duodenum. The *mesentery* is a broad, fan-shaped fold of peritoneum connecting the coils of the small intestine to the posterior abdominal wall.

Functions of the Peritoneum

1. To prevent friction as the abdominal organs move on one another and against the abdominal wall, as its free surfaces are always moist with serum and are smooth and glistening.

2. To attach the abdominal organs to the abdominal wall, except in the case of the kidneys, duodenum and pancreas, which lie behind it. The ascending and descending colon are covered only on their

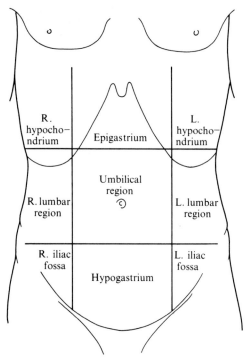

Fig. 147. Regions of the abdomen.

anterior surface by peritoneum. This means that only the transverse or sigmoid colon can be brought on to the anterior abdominal wall for a colostomy operation.

3. To carry blood vessels, lymphatics and nerves to the organs; these run between the two folds of the membrane.

4. To fight infection, as the peritoneum contains many lymphatic nodes.

Regions of the Abdomen (Fig. 147)

For the purposes of description the abdomen is divided into nine regions by two transverse and two upright lines. In describing the position of organs, the regions in which they lie may be used, e.g. the stomach lies in the left hypochondriac, epigastric and umbilical regions. The kidneys are in the right and left lumbar regions respectively. The caecum is in the right iliac fossa. The bladder rises when full into the hypogastric region.

18 The Liver, Biliary System and Pancreas

The liver is the largest gland in the body. It is situated in the upper right part of the abdominal cavity, occupying almost all of the right hypochondrium and fitting under the diaphragm. It has two main *lobes*, the right lobe being much larger than the left. The right lobe lies over the right colic flexure and the right kidney and the left lobe over the stomach.

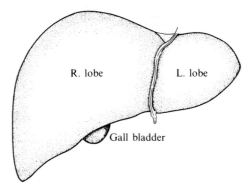

Fig. 148. The liver, general appearance.

Structure of the Liver

The liver consists of a large number of *hepatic lobules* which appear hexagonal in shape. Each is about 1 mm in diameter and has a small central intralobular vein (a tributary of the hepatic veins). Around the edges of the lobules are the *portal canals* each containing a branch of the *portal vein* (interlobular vein), a branch of the

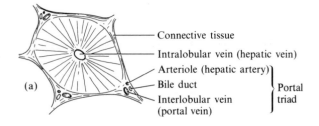

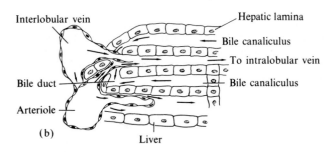

Fig. 149. (a) A liver lobule; and (b) the scheme of blood vessels and bile ducts.

hepatic artery and a small *bile duct*. These three structures together are known as the *portal triad*.

The lobules are composed of *liver cells* which are large cells with one or two nuclei and fine granular cytoplasm. The liver cells are arranged in sheets, one cell in thickness, called *hepatic laminae*. These laminae are arranged irregularly to form walls with bridges of liver cells connecting adjacent laminae. Between the laminae are spaces which contain small veins with many anastomoses between them and small bile ducts called *canaliculi*.

In the liver the portal vein brings blood rich in foodstuffs from the alimentary tract and the hepatic artery brings blood rich in oxygen from the arterial system. These divide up into smaller vessels and provide a capillary network among the liver cells forming the hepatic laminae. This capillary network then drains into the small veins in the centre of each lobule which supply the hepatic vein. These vessels carry away the blood from both the portal capillaries and the deoxygenated blood which has been brought to the liver by the hepatic arteries as oxygenated blood.

Functions of the Liver

The functions of the liver can be divided into three sections: metabolic, storage and secretory.

Metabolic Functions

1. Stored fat is broken down to provide energy. This process is called desaturation.
2. Excess amino acids are broken down and converted to urea.
3. Drugs and poisons are de-toxicated.
4. Vitamin A is synthesized from carotene.
5. The liver is the main heat-producing organ of the body.
6. The plasma proteins are synthesized.
7. Worn-out tissue cells are broken down to form uric acid and urea.
8. Excess carbohydrate is converted to fat for storage in the fat depots.
9. Prothrombin and fibrinogen are synthesized from amino acids.
10. Antibodies and antitoxins are manufactured.
11. Heparin is manufactured.

Storage Functions

1. Vitamins A and D.
2. Anti-anaemic factor.
3. Iron from the diet and from worn-out blood cells.
4. Glucose is stored as glycogen and converted back to glucose in the presence of insulin as required.

Secretory Functions

Bile is formed from constituents brought by the blood.

The Formation of Urea. The amino acids derived by the process of digestion from the protein food we eat are absorbed by the villi of the small intestine and brought by the portal vein to the liver. The amino acids required to make good the wear and tear of tissue and produce its growth are allowed to pass straight through the liver into the blood stream. Others are used to form the blood proteins. Any excess protein or second-class protein which is unsuitable for tissue building is broken up in the liver to form: (a) body fuel, composed of carbon, hydrogen and oxygen; and (b) urea, a compound containing the nitrogen present in all proteins, which is incombustible, and therefore useless unless needed for the building of tissues. This urea is a soluble substance which the blood stream carries from the liver to the kidneys for excretion from the body.

The Secretion of Bile. The bile is a thick greenish-yellow fluid secreted by the liver cells. It is alkaline in reaction. The liver secretes on average about a litre of bile per day. The bile consists of water, bile salts and bile pigments. The bile salts give the bile its alkaline reaction and include both organic and inorganic salts. Among the former is the substance cholesterol, which is the main ingredient of the typical gallstone. The bile pigments are derived from the haemo-globin of the worn-out red corpuscles which are excreted from the body by this channel, giving the normal colour to the faeces: they are also present in the blood and colour the urine.

The uses of bile are:

1. It helps to *emulsify* and *saponify* fats in the small intestine by its alkalis. In this way the surface area is increased and the action of enzymes is increased.

2. It stimulates *peristalsis* in the intestine, so that it acts as a natural aperient.

3. It is a channel for *excretion* of pigments and toxic substances from the blood stream, such as alcohol and other drugs.

4. It acts as a *deodorant* to the faeces, lessening their offensive odour. It is suggested that this may be due merely to the fact that a lack of bile means poor digestion of fat, so that fat remains in the intestine in excess, coating the other foods and preventing their di-gestion and absorption. As a result undigested protein is attacked by bacteria and, decomposing, produces an excess of sulphuretted hydrogen, the gas which causes the smell of abnormal faeces, foul drains and rotten eggs.

The Biliary System

This consists of:

1. The *right* and *left hepatic ducts* from the liver which unite to form the *common hepatic duct*.

2. The *gall bladder*, which acts as a reservoir for bile.

3. The *cystic duct*, leading from the gall bladder.

4. The *bile duct* formed by the junction of the common hepatic and cystic ducts.

The *gall bladder* is a pear-shaped organ situated on the under surface of the right lobe of the liver. From it the *cystic duct*, which is about 3 to 4 cm long, passes backwards and downwards to join the *common hepatic duct* and together they form the *bile duct*. If the bile secreted by the liver is not required immediately for purposes of digestion it passes up the cystic duct into the gall bladder where it is both stored and concentrated. The capacity of the gall bladder is between 30 and 60 ml but because of its capacity to absorb water the bile it contains becomes increasingly concentrated. When fatty food enters the duodenum the sphincter at the entrance to the bile duct

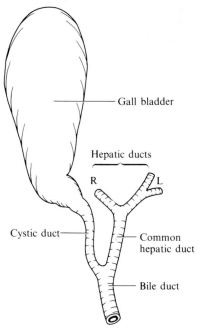

FIG. 150. The gall bladder and its ducts.

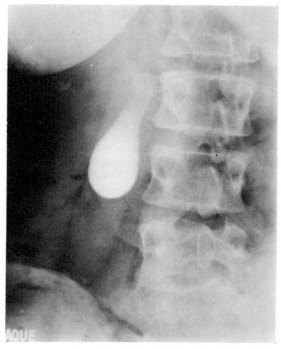

Fig. 151. An X-ray showing the normal gall bladder filled with an opaque drug (e.g. Opacol).

relaxes and bile stored in the gall bladder is driven into the intestine by contraction of the walls of the gall bladder.

From the list of its functions it will be seen that the liver is essential to life. It is, however, able to undertake more work than is usually demanded of it and a considerable part may be destroyed by disease before death occurs from liver failure.

The Pancreas

The pancreas is a soft greyish-pink gland, 12 to 15 cm in length, lying transversely across the posterior abdominal wall (see Fig. 138) behind the stomach. The *head* of the gland lies within the curve of the duodenum and the *tail* extends as far as the spleen. The *body*

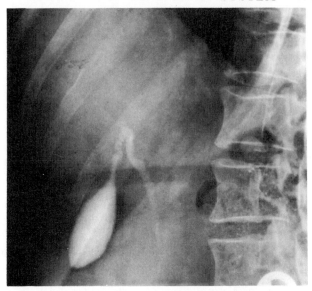

Fig. 152. The gall bladder, emptying after a fatty meal, the opaque media
showing the bile ducts.

lies between these two. The *pancreatic duct* lies within the organ. It
begins with the junction of the small ducts from the pancreatic
lobules in the tail of the pancreas and runs from left to right through
the gland, receiving ducts all the way. At the head of the pancreas
the pancreatic duct is joined by the bile duct and they usually open
together into the duodenum at the *hepato pancreatic ampulla*,
though occasionally there are two separate ducts.

The pancreas is composed of lobules, each of which consists of
one of the tiny vessels leading to the main duct and ending in a
number of alveoli. The alveoli are lined with cells which secrete
enzymes called *trypsinogen*, *amylase* and *lipase.*

Trypsinogen is converted into active *trypsin* by enterokinase, an
enzyme secreted by the small intestine. In its active form trypsin
converts peptones and proteins into amino acids.

Amylase turns starches, both cooked and uncooked, into maltose
(malt sugar).

Lipase splits fats into fatty acids and glycerol after the bile has
emulsified the fat which increases the surface area.

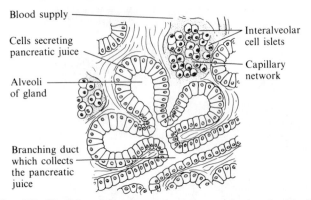

Blood supply

Cells secreting
pancreatic juice

Alveoli
of gland

Branching duct
which collects
the pancreatic
juice

Interalveolar
cell islets

Capillary
network

FIG. 153. The minute structure of the pancreas, showing the saccules
secreting the pancreatic juice and the islet cells which secre^te insulin.

Between the alveoli, collections of cells are found forming a
network in which there are many capillaries. These cells are called
interalveolar cell islets (islets of Langerhans) and they secrete a
hormone which passes directly into the blood stream. The pancreas
therefore has both a digestive and an endocrine function. Each islet
consists of two types of cell designated alpha and beta. Alpha cells
form about 25 per cent of the total number of islets and they
produce a hormone called *glucagon* which is secreted in response to a
fall in blood sugar. Glucagon stimulates the conversion of glycogen
to glucose, thus raising the blood sugar level. Beta cells form the
remaining 75 per cent of the islets and they secrete the hormone
insulin in response to a rise in blood sugar level, e.g. after a meal.
Insulin lowers the blood sugar level by stimulating the conversion of
glucose to glycogen for storage and by increasing cellular uptake of
glucose. It will be apparent therefore that the blood sugar level is
maintained by a balance between the two hormones because they
both affect carbohydrate metabolism. As the metabolism of proteins and fats are closely related to carbohydrate metabolism a
disturbance in one will affect the others.

A deficiency of insulin results in diabetes mellitus. The blood
sugar level rises above the renal threshold and glucose is lost in the
urine. Because the cells cannot utilize glucose there is an accumulation of ketone bodies as a result of the breakdown of fatty acids
which causes acidosis and may lead to coma and death if untreated.

19 Nutrition and Metabolism

Good health is dependent on satisfactory nutrition, and this is in turn dependent on plentiful supplies of the foods which are necessary to healthy life. Food is one of the essential needs of the body. Substances which can serve as food for the body are those which it can use as *fuel* for combustion, or *building material* for the repair and growth of tissues. Fuel is required to produce the energy for the activities of every living thing and to maintain the heat at which the individual exists. Building materials are necessary to repair the body tissues, since they are constantly active and being worn out by their activities; in addition, in infants and children, extra building material is required to build up the new tissues needed for the processes of growth. Fuel supplies and building materials alone are, however, not enough. Certain other substances are necessary to enable the tissues to use the building materials and the fuel supplies, and these are known as vitamins.

There are six essential foodstuffs with which the body must be constantly supplied through the foods we eat. These are:

1. Proteins
2. Carbohydrates
3. Fats
4. Water
5. Mineral salts
6. Vitamins

Every article of food contains one or more of these foodstuffs and is only of value as food because it does contain them. The proteins, water and salts are the body-building foods; the carbohydrates and fats are essentially the fuel foods, though the body can and does use protein also as fuel, if more is taken in than is required for body building, or if there is a lack of other fuel, as in starvation. The vitamins and certain salts act as regulators of tissue activity, so that, although vitamins are of no use either as fuel or as body builders, the nutrition of the tissues suffers if they are not present in sufficient quantities in the food supply, and diseases appear which can both be prevented and cured by ensuring that the vitamins are present in satisfactory quantities in the diet.

To be of use to the body these foodstuffs must be digested and absorbed. Food must therefore be of such a nature that it can be digested, i.e. broken down by digestive juices into substances that can pass into the blood stream, and be carried to the various tissues for their use. Proteins, carbohydrates and fats are complicated compounds found in plant and animal matter, and require digestion. Water and mineral salts are simple inorganic substances; they can be absorbed, therefore, without digestion and they enter into the composition of all animal and plant matter. In fact, all living matter consists largely of water. The inorganic salts, absorbed from the soil or water, are built up by living things into organic salts which are an essential part of all plants and animals.

Metabolism

Metabolism is the term applied to all the changes that occur in the body in connection with the use of food. The Greek word metabole, meaning change, provides the derivation of the word 'metabolism', which covers all the changes concerned with the use of food by the body tissues, the formation of waste products from its use, and the excretion of those waste products.

The changes included in the process of metabolism are of two distinct varieties:

1. Building-up changes, which are called *anabolic changes* or *anabolism*, e.g. the building up of muscle from the amino acids obtained from proteins, or of fat from fatty acid and glycerol.

2. Breaking-down changes, which are called *catabolic changes* or *catabolism*, e.g. the breaking down of glucose or fat into carbon dioxide and water to release energy for activity.

This division is now regarded as a little artificial as these metabolic changes are due to enzyme action and within the body cells they take place continuously and simultaneously; for example, a muscle cell constantly builds up fresh protoplasm from the amino acids provided by the blood stream; at the same time it requires energy from the burning of fuel, such as glucose, for this activity and the activity of contraction, and breaks up worn out protoplasm into waste products, such as urea, for excretion.

Metabolic changes take place all the time in every living thing, but are increased during activity, such as movement or the digestion of food, and are decreased during rest. *Basal metabolism* is the metabolism that goes on when the body is at absolute rest. The *basal metabolic rate* (BMR) is sometimes calculated as an indication of the presence of disease, since over-activity of the thyroid gland raises the rate and under-activity lowers it.

Proteins

Proteins are the most complicated of the foodstuffs. They consist of carbon, hydrogen, oxygen, nitrogen and sulphur and usually phosphorus. They are often spoken of as the nitrogenous foodstuffs, as they are the only ones which contain the element nitrogen. They are essential for the building up of living protoplasm, as this also consists of the elements carbon, hydrogen, oxygen, nitrogen and sulphur. They are found in both animal and plant matter, but the animal proteins are the most valuable to the human body as building material, since they are similar to human protein in composition. On the other hand, plant proteins are cheaper; they are more useful as body fuel than as body builders, but do provide at a lower cost some of the amino acids that the body needs for tissue building.

The *sources of protein* are:

1. Animal
   a. Eggs, containing albumin
   b. Lean meat, containing myosin
   c. Milk, containing caseinogen and lactalbumin
   d. Cheese, containing casein
2. Plant
   a. Wheat and rye, containing gluten
   b. Pulse foods (e.g. peas and beans) containing legumin

All proteins are made up of simpler substances known as *amino acids*. There are about 20 of these amino acids, but each protein contains some of these only. The amino acids are like letters from which many words can be made, each word being a different combination of letters. The protein of each different type of animal or plant is a different combination of amino acids. Ten essential amino acids are found in human protein. These are amino acids which the body cannot build up for itself. Proteins which contain all ten are called *complete proteins*, e.g. albumin, myosin, casein. Proteins which do not contain all the ten are called *incomplete proteins*, e.g. gelatin, which is contained in all fibrous tissue, and is extracted from bone and calves' feet in the making of soups and jelly. Animal proteins, such as those of eggs, milk and meat, not only contain all the ten amino acids the body needs, but contain all of them in good proportion; these are called *first-class proteins* and are the best building material for the body tissues. Plant proteins, such as gluten and legumin, contain only slight quantities of one or more of the ten amino acids essential to the body, and are therefore called *second-class proteins*, as they are not such good building material. Some first-class animal protein should always be included in the diet.

Protein Metabolism

Proteins are converted by the enzymes of the gastric, pancreatic and intestinal juices into amino acids. These are absorbed by the villi of the small intestine and carried by the portal vein to the liver.

The chief use of protein is to provide *material for body building*. The new tissue required for the purposes of growth and repair can only be made from this foodstuff, since no other foodstuff contains the nitrogen essential for the making of a living cell. The amino acids required for tissue building pass through the liver and are carried by the blood stream to all parts of the body for this purpose.

Protein can also be used as *body fuel*. Excess protein and protein unsuitable for body building, such as the second-class protein obtained from plant foods, are split up in the liver to form:

1. *Body fuel* in the form of glucose (containing carbon, hydrogen and oxygen)
2. *Urea* or nitrogenous waste matter (containing nitrogen, which is incombustible, and hydrogen)

This process is termed the de-amination of the amino acids. The nitrogenous content of the amino acids not required for body building is converted first into ammonia which, for the most part, combines with carbonic acid and splits up into urea and water in the liver.

The glucose is either burnt or stored as required. The urea is readily soluble, and, being useless for fuel, is carried away by the blood stream and excreted from the blood by the kidneys.

Protein will also serve as fuel in conditions of starvation. Combustion must go on continuously to maintain life. If there is lack of other fuel, protein will be sacrificed for use as fuel, even though it is needed for processes of repair and growth. As a result, weakness and loss of weight will follow when either the body cannot obtain food, as in famine, or cannot digest and absorb it, as in disease.

The *waste products* of protein metabolism are *urea* and to a lesser extent *uric acid* and *creatinine*. Uric acid is less soluble than urea, and comes particularly from the nuclear material in our food. Creatinine is the waste product of the breaking down of our own body protein. All these protein wastes are excreted by the kidneys in the urine. About 30 g of urea leave the body each day in this way, as also do traces of uric acid and creatinine.

Carbohydrates

Carbohydrates include *sugar* and *starch*. They consist of carbon, hydrogen and oxygen, and always contain hydrogen and oxygen in the same proportions as in water, i.e. twice as much hydrogen as oxygen. They are the chief sources of body fuel, being most easy to digest and absorb and most readily burnt in the tissues, being broken down into carbon dioxide and water. They are obtained chiefly from plant foods.

The *sources of starch* are:

1. Cereals, e.g. wheat, rice, barley
2. Tubers and roots, e.g. potatoes and parsnips
3. Pulse foods, e.g. peas, beans and lentils

The *sources of sugar* are:

1. Sugar cane (sucrose)
2. Beetroot and all sweet vegetables and fruits, e.g. grape sugar or glucose.
3. Honey
4. Milk, containing lactose or milk sugar

Sugars are of three types:

1. Simple sugars (monosaccharides), such as glucose or grape sugar (formula $C_6H_{12}O_6$)
2. Complex sugars (disaccharides), such as cane sugar or sucrose, and milk sugar or lactose (formula $C_{12}H_{22}O_{11}$)
3. Polysaccharides, the most complex carbohydrates; starches such as potatoes, cereals and root vegetables

Starch differs from sugar in that it is insoluble in water. Plants store sugar in the form of starch to prevent it from escaping in solution into the water in the soil in which they live. (Formula for starch $n(C_6H_{10}O_5)$, a polysaccharide; n stands for different numbers in the different starches of various plants.) All carbohydrates are reduced to monosaccharides before they can be absorbed from the digestive tract.

Carbohydrate Metabolism

Carbohydrates in the form of sugar and starch are acted on by the enzymes of the saliva, pancreatic and intestinal juices, and converted into simple sugars such as glucose ($C_6H_{12}O_6$). This simple sugar is absorbed by the villi of the small intestine, passes into the blood capillaries, and is carried by the portal vein to the liver where excess sugar is stored as glycogen.

The use of glucose is to serve as the main *body fuel* for the production of energy for work and heat. The glucose required for immediate use is carried straight through the liver into the hepatic veins and inferior vena cava, and so enters the circulation. Any excess not required for the body's immediate needs is converted by the liver cells into *glycogen*, which is insoluble and is stored in the liver until required for use. Glycogen in the animal world is similar to starch in the plant world. Both are insoluble and are formed from sugar by the splitting-off of water from it. When sugar is needed by the body the glycogen is converted back into glucose, which the body fluids dissolve, so that it passes into the blood stream. Both the formation of glycogen from glucose and the forming of glucose from glycogen are the work of enzymes produced by the liver cells.

Glucose is specially required by the most active tissues of the body, the muscles and glands, but all tissues need it to some extent. The muscles, like the liver, are able to store it to a slight extent in the form of glycogen. The tissues of the body utilize their fuel very economically. Glucose is burnt to produce energy for muscle contraction, but more energy is released by complete combustion than is required for the work of the muscles, and the excess energy is used to build up glycogen again from some of the partially burnt fuel. Actually it is estimated that only about one-fifth of the fuel is completely burnt down into the waste products carbon dioxide and water. The remaining four-fifths, after partial combustion, is built up again into glycogen ready for use when the muscles require it. The energy for this building-up work is obtained from the complete combustion of the other fifth of the fuel.

The *waste products* of the combustion of the carbohydrates are *carbon dioxide* and *water*, which are carried away by the blood stream and excreted from the body. The lungs excrete the carbon dioxide and water, which is also got rid of by the skin and kidneys. Combustion of glucose may be incomplete if there is deficient oxygen supply, and gives rise to acid bodies which cause the pain of acute cramp, as occurs in violent exercise in the muscles or heart wall. The acids cause the blood vessels to dilate, which increases the blood supply, and thus the oxygen supply, and results in the passing off of the pain, as combustion can then be completed.

The metabolism of carbohydrate is controlled by *insulin*, the internal secretion of the pancreas (see Chapter 18). Without insulin the tissues are not able to burn glucose nor the liver to store it as glycogen. If the insulin supply is normal the amount of glucose in the blood varies very slightly and is normally about 4.4–6.7 mmol/l of blood. After a meal it will rise slightly; the rise immediately stimulates the pancreas to make more insulin, and the extra insulin causes the liver and muscles to store any excess quickly and the blood sugar falls to normal again. If there is deficiency of insulin the blood sugar rises too high and neither liver nor muscles can store glucose to the normal extent. This occurs in diabetes, a condition in which the pancreas is diseased and fails to produce the normal quantity of insulin, so that the blood contains excess sugar, and sugar is excreted through the kidney into the urine. Fat is burnt instead of sugar and this is dangerous (see page 226). On the other hand, too much insulin in the blood causes excess glucose to be stored and leaves insufficient in the blood stream. This is very serious and may even cause convulsions, coma and death. It is chiefly seen in the giving of an overdose of insulin.

Fats

Fats, like carbohydrates, consist of carbon, hydrogen and oxygen, but do not contain as much oxygen in proportion to the hydrogen present. They also serve as body fuel. They are the best source of fuel from the point of view that 1 g of fat produces twice as much energy as 1 g of sugar. On the other hand, they are not so easy to digest and absorb and not so readily burnt in the tissues. Fats can only be completely burnt to form the final products of combustion, carbon dioxide and water, if they are burnt with sugar. If sufficient sugar is not burnt with them, the combustion of fat is incomplete and acid or acetone bodies are formed in the tissues. These acetone bodies cause fatigue in the muscles and if present in large quantities alter the reaction of the blood, causing a condition known as acidosis which may lead to coma and death. Severe acidosis is only likely to occur in diabetic patients, who are unable to burn sugar, and in starvation, when the small quantity of sugar which the body can store has been used up and the large quantities of fat which the body can store are forming the main source of body fuel.

Fats are obtained from both animal and plant matter. The chief sources of fat are:

1. Animal
   a. Fat meat and fish oils
   b. Butter
   c. Milk and cream
2. Plant
   a. Nut oil, contained in margarine
   b. Olive oil

Fats are compounds of glycerin and fatty acids. Different fats contain different acids, e.g. fat meat contains stearic acid, butter contains butyric acid, olive oil contains oleic acid, and so on.

Animal fats are more expensive than plant fats but are more valuable as food because they contain the fat-soluble vitamins A and D provided the animals have been exposed to sunlight and not kept in stalls. However, plant fat can be made equally valuable by exposing it to the action of ultraviolet rays, and all margarine, whether manufactured from fish oil or plant oil, is so treated today to ensure the supply of these vitamins. In this way the population has been protected from deficiency diseases due to lack of these fat-soluble vitamins.

Fat Metabolism

Fat is emulsified by alkalis and split up into fatty acids and glycerol by the lipase of the pancreatic and intestinal juices. These substances are absorbed by the lacteals and pass through the lymphatic circulation, up the thoracic duct and into the blood stream.

Fat serves as *fuel* for the production of heat and energy in the tissues. It is a better fuel than glucose in that it produces twice as much heat and energy per gramme of fuel used. On the other hand, it is less easy to digest and absorb and less satisfactory to burn.

Fats, before they can be used as body fuel, must be prepared by the liver for combustion in the tissues. This again is a chemical process carried out by the liver cells, and is known as the desaturation of fats.

Fat is also required for the building of various tissues, e.g. nerve tissue, fatty tissue and marrow; fat derivatives are found in the secretions of certain glands.

Fat not required for immediate use as fuel can be stored as fatty tissue. This is found particularly in the subcutaneous tissue and in the body cavities, but whereas sugar can only be stored in very limited quantities, the total being about 225 g in liver and muscles, fat can be stored in very large quantities, and may be found in large amounts even in the muscles themselves. This is, however, undesirable, as it causes increase in weight, which limits activity, setting up a vicious circle.

The putting-on of fat does not necessarily mean that too much fat is being eaten, as the body can convert excess glucose into fat for storage and excess protein not required for body building into glucose. The metabolism of the three foodstuffs is therefore closely linked and the taking of food of any kind in excess of the needs of the tissues will lead to an increase in weight.

The *waste products* of fat metabolism are *carbon dioxide* and *water* if combustion is complete; these are excreted by the lungs, skin and kidneys, as in the case of carbohydrates. If combustion is incomplete, the *acetone bodies* formed also leave the body by the same routes. The volatile acetone can be smelt in the 'sweet' breath, and acetone and diacetic acid can be found in the urine.

Water

Water is a simple compound of hydrogen, 2 parts, and oxygen, 1 part. It forms two-thirds of the body and is present in most of the foods we eat. Lean meat is three-quarters water, milk contains 87 per cent water, while cabbage contains as much as 92 per cent water. In addition to the water contained in food the body needs 2 to 3 litres of water every day. Water is required for many purposes, the chief being:

1. The building of body tissues and body fluids
2. The excretion of waste products
3. The making of digestive and lubricating fluids
4. The cooling of the body by the evaporation of sweat

Water Balance

Water is one of the essential foodstuffs, but is a simple substance which can be absorbed and used in the body without chemical change. It enters into many of the metabolic changes which occur in the body, combining with proteins, carbohydrates and fats in digestion and being split off from them when they are used as fuel to produce energy. It is general today to consider the balance rather than the metabolism of water and salts, and the electrolytes which the salts form in the body.

Water forms the greater part of the body cells and the body fluids. Roughly two-thirds, or more accurately 60 per cent, of the body weight consists of water. This proportion must be maintained

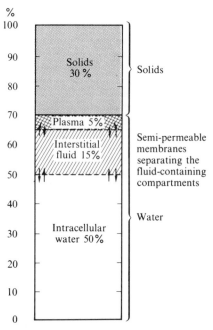

FIG. 154. Diagram to show the average percentage proportions by weight of solids and water, and the distribution of water in the body. The arrows indicate the passage of water and solutes between the three compartments containing them.

since all complicated processes of life demand its presence. Of this water 70 per cent is inside the body cells (intracellular) and the remaining 30 per cent is in the body fluids (extracellular); 15 to 20 per cent being in the interstitial spaces in the tissues, bathing all the body cells, even the bone cells, and the remaining 10 to 15 per cent forming the fluid of the blood, i.e. the plasma and the lymph. These three volumes of fluid are separated only by thin semi-permeable membranes, the cell walls and the capillary walls; water constantly passes through these walls from one of these areas to the others,

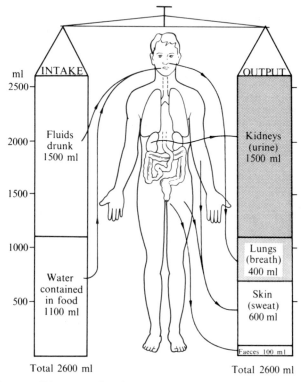

FIG. 155. Diagram to show how water balance is maintained by various organs.

though the volume of each remains remarkably constant in normal health.

The water of which our bodies are largely made is, however, not static. Fresh water is taken in by the body each day and passes out of it by a number of channels. The total quantities taken in and passed out must balance one another. Water is taken in as water and other fluids drunk, and also in the foods eaten, which, like the body, consist largely of water. On an average the healthy man takes in 1.5 litres of fluid as water and other drinks daily, and a little over one litre in his food, an average of 2600 to 2800 ml. A similar quantity is passed out of the body by the lungs as water vapour (400 to 500 ml), by the skin as sweat (500 to 600 ml), by the kidneys as urine (1000 to 1500 ml) and in the faeces a small quantity (100 to 150 ml).

The quantities lost in urine, sweat and water vapour from the lungs vary with conditions. In hot weather and heavy work more sweat is produced to cool the body, and less urine is passed; at the same time the loss of fluid causes thirst, so that more fluid is drunk. In fever the same changes take place; there is increased loss of fluid and there must be increased intake to balance it.

Electrolytes

In addition to the balance of the quantity of water in the body, the body fluids must be of the right reaction, i.e. they must contain the correct balance of electrolytes. These electrolytes are minute particles split off from the molecules of the various salts and dissolved in water. They are called ions (see page 8). They carry electric charges and are of two types, negatively charged particles (anions) and positively charged particles (cations). The total of negatively charged ions must balance the total of positively charged ions. The chief negatively charged ions are chloride (Cl), bicarbonate (HCO_3), and phosphate (PO_4); chloride and bicarbonate are present in quantity in plasma and interstitial fluid, while intracellular fluid contains mainly phosphate. The positively charged ions are sodium (Na) and potassium (K), with a little calcium (Ca) and magnesium (Mg). Sodium is present as the main positive ion in plasma and interstitial fluid, while intracellular fluid contains chiefly potassium.

Salts and the ions produced from them are constantly taken into and lost from the blood. Salts are contained in all foodstuffs, but added salt, used in cooking and in eating, is necessary to maintain the normal electrolyte balance and replace the salt lost in urine and sweat; 3 to 4 g of sodium chloride, or common salt, is needed daily.

The quantities of the various salts are estimated as millimoles/litre (mmol/l) of the ions they produce. In plasma there are normally 155 mmol/l of negatively charged ions, balanced by 155 mmol/l of positively charged ions. The bulk of these are provided by chloride (102 mmol/l) and sodium (143 mmol/l). These figures are given for reference, as the nurse will see them in laboratory reports to check for deficiencies or excess of sodium, potassium chloride, bicarbonate, etc., in the blood. These may result from deficiency in intake or excessive or reduced loss, especially in abnormal conditions affecting the kidney, or from heavy sweating or fever.

Acid–Base Balance

The acid–base balance of the body fluids affects their reaction. Normally the body fluids are slightly alkaline and vary very little during life, since this reaction must be correct for the action of the various enzymes which control the cell activities, such as digestion, the use of food for growth and energy production, and the production and excretion of waste products. Normally venous blood is a little less alkaline than arterial blood because of the carbon dioxide and other acids produced in the tissues. The interstitial fluid and the intracellular fluid are similar but slightly more alkaline. This balance is maintained by the 'buffers' these fluids contain. Alkalis and proteins neutralize acids produced by tissue activities, preventing acidosis, or ketosis, a condition which leads to coma and death if the body fluids become less alkaline than normal. (The normal average is pH 7·4; see Chapter 1.) In the same way, acids such as carbonate and chloride neutralize excess alkalis in the body fluids, preventing alkalosis, a condition which can prove fatal if it is not corrected. It may be caused by taking in an excess of alkaline salts such as sodium bicarbonate, or by loss of acid from excessive vomiting, or from continuous gastric aspiration after operation, or in respiratory conditions, or from retention of normal quantities of an alkali such as potassium, through renal failure.

Mineral Salts

Salts are produced by the action of an acid on a mineral, which is termed the base of the salt, e.g. sodium chloride or common salt, produced by the action of hydrochloric acid on sodium; calcium lactate, produced by the action of lactic acid on calcium. Various salts and the ions split off from them are required by the body, every tissue and fluid in the body containing them. Chlorides, carbonates and phosphates of sodium, potassium and calcium are particularly important. Salts are required for body building and are also regulators of tissue activity. They provide the electrolytes, capable of carrying positive and negative electric charges in the body fluids (see Chapter 1). A correct balance of electrolytes is essential for normal activity in the body tissues and fluids.

Sodium is present in all tissues; the body contains it in the form of sodium chloride to the extent of 9 g to 1 litre (0·9 per cent). It is present in the same concentration in all tissue fluid. Sodium carbonate and sodium phosphate are also always present in the blood and tissues. The carbonates give the alkaline reaction to the blood and form the alkali reserve, neutralizing the carbonic acid produced by fuel combustion. The phosphate is the carrier of the acids produced by the breaking down of the body building foods and carries them to the kidneys, by which they are excreted. It is obtained from our food, particularly animal foodstuffs, and from the rock salt used in cooking.

Potassium is present in all tissue cells, where it replaces the sodium of blood and tissue fluid as the base and the source of the positively charged ions. It is obtained from our food, particularly plant foodstuffs.

Calcium is present in all tissues, particularly in bone, in teeth, and in blood, and is necessary for normal functioning of nerves. It is obtained chiefly from milk, cheese, eggs and green vegetables, and to some extent from hard water, though this inorganic form is less valuable than the organic salts in plant and animal foods. Adults require 400 to 500 mg daily.

Iron is essential for the formation of the haemoglobin of red blood cells. It is obtained from green vegetables, particularly spinach and cabbage, egg yolk and red meats. Men require 10 mg daily; women 10 to 15 mg.

Phosphorus is also needed for the building of the body tissues. It is obtained from the yolk of egg, milk and green vegetables.

Iodine is required for the formation of the secretion of the thyroid gland. It is obtained from seafood and is present in green vegetables, which have derived it from the fine spray of sea water in the air which is carried inland by winds.

Calcium, iron and iodine are the only minerals likely to be insufficient. The others are present in adequate amounts in the diet.

Vitamins

Vitamins are substances which are also essential to normal health, although they are of no value to the body as fuel nor as building material. Without them diseases occur which are known as deficiency diseases. They are present in small quantities in living foodstuffs and are required in minute traces only each day. Their discovery dates back to the period following the 1914–18 war, and as their composition was at first unknown they were named after the letters of the alphabet. They can today be made synthetically to a large extent.

There are numbers of vitamins now identified, but experimental work still continues. The chief vitamins are: vitamin A, the vitamin B complex, vitamin C, vitamin D, vitamin E and vitamin K.

Vitamin A. Vitamin A is present in animal fats, being fat-soluble. In carrots and green vegetables, and in all yellow fruits a substance called *carotene* is found, which, like the green colouring matter chlorophyll, is formed under the influence of sunlight. This is a precursor of vitamin A, and animals, including man, can turn the carotene present in their food into vitamin A within their bodies. Lack of vitamin A causes stunted growth and lowered resistance to infection. The mucous membranes particularly are unhealthy and an easy prey to bacteria when the diet is deficient in it. The conjunctiva of the eye is affected and a form of conjunctivitis occurs, known as *xerophthalmia*, in which the conjunctiva loses its transparency, and a horniness or cornification develops in it. The retina is also affected and night blindness (i.e. inability to see at night) develops—a symptom which became apparent in the 'black-out' during the war, when some people found they could not see as well as others in the dark, because their diet had not been satisfactory.

Vitamin B. The vitamin B complex consists of a number of factors, although originally thought to be a single substance. These factors were confused because they had a similar distribution, being found particularly in the husks and germs of cereals and pulses, and in yeast and yeast extracts. They are also present to a lesser extent in vegetables, fruit, milk, eggs and meat. White flour and the bread, cakes and pastries made from it do not contain these factors, nor do polished rice and barley. Hence brown bread and wholemeal flour have a food value that white bread and flour lack and where these latter form a large part of the diet a subnormal state of health may be present, or even deficiency disease, as occurred in many of the prison camps during the last war, especially in the Far East. The chief factors in the vitamin B complex are:

Vitamin B (aneurine or thiamine). This is essential for carbohydrate metabolism and controls the nutrition of nerve cells; a marked deficiency of it leads to beri-beri, in which there is inflammation of the nerves, causing paralysis and a loss of tone and activity in the intestinal muscles, with constipation, while the patient complains of loss of appetite and a burning sensation in the feet.

Vitamin B_2 (riboflavine). This is essential for proper functioning of cell enzymes.

Vitamin B_3 (nicotinic acid). This is essential for carbohydrate metabolism. A lack of this causes pellagra, which leads to skin eruptions, gastro-intestinal changes and mental changes.

Vitamin B_6 (pyridoxine). This is believed to be necessary for protein metabolism.

Vitamin B_{12} (cyanocobalamin). This is the anti-anaemic substance or factor absorbed by the villi of the small intestine and stored in the liver. It is satisfactorily absorbed only in the presence of an intrinsic factor produced by the lining of the stomach and hydrochloric acid. Vitamin B_{12} is essential for the proper development of red cells in the red bone marrow. Lack of it, or of its absorption, causes pernicious anaemia.

Folic acid. This is part of the vitamin B complex. It is also necessary for the maturation of red blood cells.

Vitamin C. Vitamin C (ascorbic acid) is water-soluble and is found in fresh fruit, particularly citrus fruits (oranges, grapefruit and lemons) and in green vegetables and potatoes. It is important in tissue respiratory activity, wound repair and resistance to infection, and affects the condition of capillary walls, which become abnormally fragile if it is not present in plentiful supply in the diet. Lack of it causes scurvy, so that it is called the antiscorbutic vitamin. It is particularly readily destroyed by heat, hence some fresh fruit or salad should be included in the daily diet. Cabbage and other greens provide a very rich source of vitamin C and are comparatively cheap.

Eaten raw, as salads, in a finely shredded form, they are very palatable and most valuable, provided the cabbage is fresh and crisp. Cooking for a long period of time lessens the content of vitamin C, and for this reason cabbage should be finely shredded and cooked in boiling water only until tender, i.e. for 10 to 15 minutes.

Under normal circumstances the eating of a good mixed diet, containing plenty of fresh fruit and vegetables, makes the taking of vitamin C in tablet form unnecessary and inadvisable.

Vitamin D. Vitamin D is fat-soluble and is found with vitamin A in animal fats, provided the animal has been in the sun. Cod-liver and halibut-liver oil are very rich in it. Halibut oil is more expensive as the supplies are less plentiful, but it is richer in content, so that it is useful for those who cannot digest cod-liver oil. Vitamin D can also be made in the skin by the action of ultraviolet rays on the ergosterols present. Further, it can be manufactured by subjecting the sterols in fats to the action of ultraviolet rays. The product of this process is named *calciferol* and this can be obtained in tablet form. It is identical with the vitamin D present in animal fat and can be used in illness and in emergencies as a substitute for it. It is essential for the development of bone and teeth, affecting the use of calcium and phosphorus in the body. Lack of vitamin D causes rickets, so that it is known as the antirachitic vitamin. This disease was common among the poor of industrial areas who ate margarine in place of butter, especially in countries like the British Isles, where there was not a great deal of sunshine. However, modern methods of manufacturing the vitamin have made it possible to add it to margarine, with the result that rickets as a disease has practically disappeared in this country, since it has become compulsory for all makers of margarine to ensure that it contains the vitamin.

Vitamin E. Vitamin E is present in vegetable oils. It is present in cereals and has been shown to be essential for reproduction in rats. Little is known of its importance in human beings.

Vitamin K. Vitamin K is fat-soluble and can be obtained from green vegetables and liver. Persons who have had deficiency in their diet show a tendency to haemorrhage, since vitamin K helps to form the prothrombin in the blood. It is synthesized in the intestines by bacterial action.

The quantities of the various vitamins required in the normal diet are estimated in milligrams; pure vitamins can be given in measured quantities on the doctor's prescription.

Roughage

The food should also contain cellulose. This is indigestible and therefore remains in the bowel and stimulates it to empty itself. It is called roughage. Cellulose forms the fibrous part of plant foods and is present in all green vegetables, in fruits, the skin of beans and peas, and in wholemeal flour and bread. These foods are therefore essential to prevent constipation.

Diet

Diet is the daily ration of foods required by the individual. Man requires a mixed diet, i.e. a diet consisting of different animal and plant foods, as no single article of food contains all the essential foodstuffs in the proportions required for health. Milk and oysters contain all the nutrients or foodstuffs, but not in the proportions the body requires. Cow's milk contains on an average:

Protein (caseinogen, lactalbumin)	4%
Lactose	4–5%
Fats	3·5%
Mineral salts	0·7%
Water	87–88%

A well-balanced diet should contain the foodstuffs roughly in the following proportions: 1 part protein, 1 part fat to 4 parts carbohydrate. In addition it must contain traces of the various vitamins. The standard requirements per day are estimated by many authorities to be:

Protein	46–56 g
Fats	66–80 g or a quarter of total calorie intake
Carbohydrate	300–400 g or 50–60% of total calorie intake

It is now accepted that health can be maintained on a diet containing 50 g of protein and 70 g of fat. These are the scarcer and more expensive foodstuffs and are derived largely from animals. Larger amounts of protein are required by persons carrying out heavy work, either manual labour or a hard athletic programme; also by pregnant women and growing children, especially in late adolescence. In the first group there is a quicker breaking down of tissue, and in the second there is the demand for building material to make new tissue for growth, in addition to the constant repair of cells that are worn out by the normal activities of life. More fat is especially desirable for persons living in a cold environment and when prolonged staying power is required, since it is more slowly digested and absorbed. Where a high calorie intake is necessary, about half the calories should be derived from fat, otherwise the meals become too bulky. In many of the under-developed and thickly populated countries, lack of animal protein and fat is responsible for a low standard of health and a high death rate, especially in the lower age groups.

The amount of carbohydrate required depends on the energy output. A man doing heavy work requires more than a sedentary worker.

Caloric Value of Diet

The value of foods is reckoned by the amount of heat which they yield on combustion. The heat is measured in Calories, now known as kilocalories. A Calorie is the amount of heat required to raise the temperature of 1 litre of water 1°C. The Caloric value of each of the foodstuffs is known and is as follows:

1 g of protein has a heat value of 4 Calories.
1 g of carbohydrate has a heat value of 4 Calories.
1 g of fat has a heat value of 9 Calories.

The Caloric value of an average diet should be 2500 to 3000 Calories per day. A man doing active work will require 3300 Calories. A sedentary worker will require about 2850 Calories. A woman will require about 2200 Calories.

The Caloric value required by the individual is affected by:

1. Age. Children require more in proportion to their weight, because they need food for growth as well as other activities; a baby requires 50 Calories per pound body weight. They need more protein in proportion to fuel foods, because of their rapid growth.

2. Exercise.

3. Sex. A male requires more than a female, the female requiring four-fifths of the quantity the male requires. In the 16 to 18 year

age group the male adolescent requires on an average one-fifth more than the male adult and the female adolescent the same quantity as the male adult; this is because of the rapid growth and development that takes place at this period, in addition to the energy expended in physical activity.

4. Weight and build. The heavier the individual, the more food is required, except where weight is due to fat.

5. Climate and weather. In hot climates and hot weather a diet of lower Caloric value is sufficient.

6. Temperament. The placid person requires less food than an excitable one.

The energy value of food can also be measured in kilojoules. One Calorie is equal to 4.2 kilojoules.

20 Endocrine Glands

The endocrine glands are organs producing secretions called *hormones* which are poured directly into the blood stream from the glandular cells. It is for this reason that they are also known as *ductless glands*.

Hormones are organic compounds manufactured by the glands from substances carried in the blood. They are mainly protein derivatives, but some are steroids. They are then carried by the blood stream to other parts of the body where they have a specific effect.

The endocrine glands have been investigated and are described as though they are separate entities but in fact their functions are closely interrelated. Initially the functions of the hormones were deduced by observing the effects of disease, destruction or overgrowth of glands. In recent years hormones have been isolated, obtained in pure form, analysed and in some cases successfully synthesized. The glands secrete their hormones continuously but the quantity of secretion can be increased or decreased according to body needs. Control of hormone secretion occurs in several different ways:

1. Nerve cells produce chemical substances which are carried to the gland and cause secretion.

2. The gland responds to impulses from the autonomic nervous system.

3. One gland produces a hormone which affects a second gland. The second gland produces its hormone, which influences secretion of the first gland. This is called a 'feed-back' mechanism.

4. The gland responds to blood levels of substances other than hormones.

The Hypophysis

The hypophysis, or pituitary gland, lies in the hypophyseal fossa of the sphenoid bone in the base of the skull. It is attached by a neural stalk to the *optic chiasma* at the base of the brain.

The gland consists of an *anterior lobe*, or *adenohypophysis* and a *posterior*, or *neural*, lobe. The anterior lobe is an endocrine gland in the true sense, while the posterior lobe is derived from the brain and consists of nervous tissue; it is connected directly to the *hypothalamus*. The two lobes are essentially two different endocrine glands and are commonly called the anterior and posterior pituitary glands.

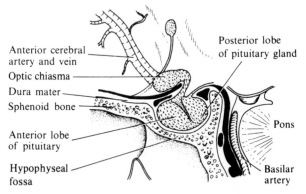

Anterior cerebral artery and vein

Optic chiasma

Dura mater

Sphenoid bone

Anterior lobe of pituitary

Hypophyseal fossa

Posterior lobe of pituitary gland

Pons

Basilar artery

FIG. 156. The pituitary gland, shown in section, lying in the sella turcica or hypophyseal fossa. Note the stalk joining it to the base of the brain.

The anterior lobe of the hypophysis is sometimes referred to as the 'master gland' of the endocrine system because it has an important influence in regulating the function of other glands. However, the glands really act in concert, one becoming active if the others are not producing sufficiently and output of hormones being decreased when other glands are active.

The anterior lobe produces a number of hormones:

1. *Thyroid-stimulating hormone* (*TSH or thyrotrophic hormone*) influences all aspects of thyroid gland function including stimulation of the accumulation of iodine in the gland for conversion into thyroid hormones, manufacture of the hormones and their release

into the circulatory system. Through its action on the thyroid gland TSH is involved in the regulation of metabolic rate, breakdown of fat and the increase of the water content of certain tissues.

2. *Adrenocorticotrophic hormone (ACTH)* regulates the development, maintenance and secretion of the cortex of the suprarenal glands. General metabolic effects of this hormone include mobilization of fats, production of hypoglycaemia and increase in muscle glycogen. It is also involved in body resistance to stress.

3. *Somatotrophic (growth) hormone* exerts its influence mainly on the hard tissues of the body, though there is some effect on the soft tissues. The hormone increases the rate of growth and maintains size once maturity has been reached. It controls the rate of growth in epiphyses and other ossification centres of the skeletal system. Oversecretion of this hormone causes overgrowth of the long bones in children (gigantism) and acromegaly in adults. In acromegaly the bones cannot increase in length because the epiphyseal plates have closed so the bones become thicker and coarser; the lower jaw, hands and feet are particularly affected. Undersecretion of the growth hormone produces dwarfism. People who are very short or very tall, due to under-secretion or over-secretion of this hormone, are usually of normal intelligence, unlike the person who suffers from under-secretion of the thyroid gland. Metabolic effects of this hormone include an increase in the conversion of carbohydrates to amino acids and in the uptake of amino acids by the cells; mobilization of fat from the storage areas and increased fat metabolism and an increased blood sugar level.

4. *Follicle-stimulating hormone (FSH)* controls the maturation of ovarian follicles in the female and the production of sperm in the male.

5. *Luteinizing hormone (LH)* causes changes in the female which lead to the formation of the corpus luteum; it also helps to prepare the breasts for the secretion of milk. In the male a comparable hormone is called *interstitial cell stimulating hormone (ICSH)* which acts on the testes and controls the secretion of the male sex hormone, testosterone.

6. *Lactogenic hormone (prolactin)* is one of several hormones involved in the production of milk by the breasts and appears to function only in females.

The posterior lobe of the hypophysis releases two hormones, but it is important to note that these are produced in the hypothalamus and only stored in, and released from, the posterior lobe.

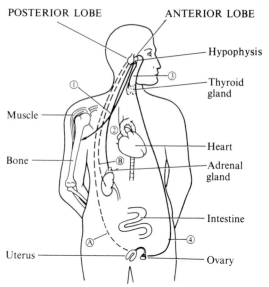

POSTERIOR LOBE   ANTERIOR LOBE

Hypophysis

Thyroid gland

Muscle

Heart

Bone

Adrenal gland

Intestine

Uterus

Ovary

Fig. 157. Diagram to show the effects of the pituitary hormones. The figures and letters refer to Table 15.

1. *Oxytocin* exerts its effects mainly on the unstriped muscle of the pregnant uterus and the cells around the ducts of the breasts, although it does also promote a generalized contraction of unstriped muscle throughout the body.

2. *Antidiuretic hormone (vasopressin)* causes an increase in the reabsorption of water by the kidney tubules so that less urine is excreted. Under-secretion of ADH causes less water to be reabsorbed and excessive amounts of very dilute urine are excreted—a condition known as *diabetes insipidus*. The hormone also causes a certain amount of vasoconstriction with consequent rise in blood pressure, but in man this occurs mainly in the coronary vessels.

The Thyroid Gland

The thyroid gland is situated at the front and sides of the neck, opposite the lower cervical and first thoracic vertebrae. It consists of

Table 15

Action of Pituitary Hormones

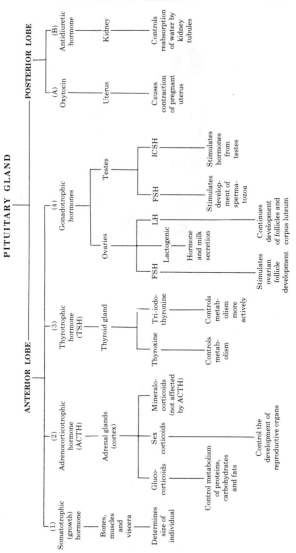

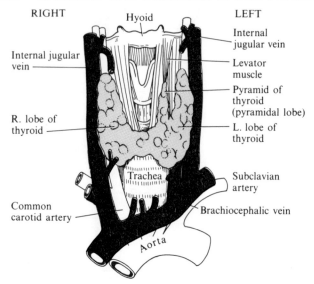

FIG. 158. The thyroid gland.

two lobes, one on either side, joined by a narrower portion called the isthmus which crosses in front of the trachea just below the larynx. The gland is formed of a number of closed *follicles* containing a yellow semi-fluid material called *colloid*.

The cells produce a hormone called *thyroxine* which may be released directly into the blood stream if it is required or may be linked to the protein substance, *thyroglobulin*, and stored in the colloid. The amino acid *tyrosine* and the mineral iodine are both essential for the formation of thyroxine.

The function of thyroxine is to regulate metabolism in the tissues. Together with growth hormone it ensures proper development of the brain; it increases urine production, protein breakdown and the uptake of glucose by the cells. A second hormone, tri-iodothyronine, has a more immediate, though similar, effect to thyroxine.

Under-secretion of thyroid hormones in a child produces the condition of *thyroid cretinism* which if untreated results in a mentally retarded dwarf. In an adult, under-secretion produces the condition known as *myxoedema*. In both these conditions the skin is

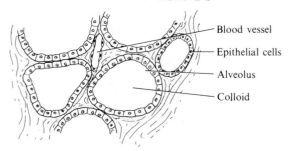

FIG. 159. Thyroid follicles.

dry and coarse, and the hair dry, coarse and lank. The metabolic rate is lowered so the patient is obese, has a low body temperature and feels the cold. Over-secretion of the hormones produces *thyrotoxicosis*, a condition in which there is an increase in the metabolic rate. The individual is anxious and nervous and has a fast pulse rate. The skin is fine and moist, the patient feels the heat and loses weight in spite of having a good appetite. He may or may not have protruding eyeballs, a condition known as exophthalmos.

Any enlargement of the gland is known as *goitre* and this may be present with or without hyperthyroidism. The presence of a goitre may cause pressure on the trachea or on the recurrent laryngeal nerve causing hoarseness.

The Parathyroid Glands

The parathyroid glands usually lie between the posterior borders of the lobes of the thyroid gland and its capsule. They are about the size of a pea and there are usually four—two behind each lobe. This number occasionally varies.

The glands produce a hormone called *parathormone* which regulates the distribution and metabolism of calcium and phosphorus in the body. Over-secretion causes calcium to be moved from the bones into the blood, from which it is excreted in the urine. The bones become porous and brittle and the increased level of blood calcium may cause the formation of kidney stones (renal calculi). Under-secretion of the hormone causes low blood calcium levels resulting in the muscular rigidity and spasm seen in the condition known as *tetany*. This complication may occasionally be seen after surgery on the thyroid gland in which the parathyroids have inadvertently been removed. Calcium is given to correct the condition.

The Suprarenal Glands

The suprarenal (or adrenal) glands lie one above and in front of the upper end of each kidney, behind the peritoneum. They are surrounded by areolar tissue containing a considerable amount of fat. Each gland consists of two quite separate endocrine glands: the outer part called the cortex and the inner part called the medulla.

Cortex

The cortex is subdivided into three zones: the outer zone produces the mineralocorticoids, the central zone produces the glucocorticoids, the inner zone produces sex hormones.

Mineralocorticoids are the steroids that regulate electrolyte metabolism. *Aldosterone* is an important mineralocorticoid and it has a regulatory effect on the relative concentrations of minerals in the body fluids, particularly sodium and potassium. It therefore also affects the water content of the tissues. When there is a deficiency of this hormone there is increased excretion of sodium and chloride ions, too much water is lost from the body in the urine and there is a corresponding fall in the concentrations of sodium, chloride and bicarbonate in the blood, resulting in a lowered pH (or acidosis).

Glucocorticoids are important in carbohydrate metabolism. They increase the conversion of protein to glycogen for storage by the liver (gluconeogenesis) and decrease the utilization of glucose by the cells, thus increasing the blood sugar level. *Cortisone*, *cortisol* (hydrocortisone) and *corticosterone* are the primary glucocorticoids and they may be given to people who have chronic inflammatory and allergic responses, such as in rheumatoid arthritis, because the steroids stimulate the anti-inflammatory and repair mechanisms of the body. Chronic stress of any kind produces increased production of the glucocorticoids which in turn help the body to resist stress; however, continuing high levels of glucocorticoids may increase the likelihood of ulcer formation, increase the blood pressure and lower the body's resistance to infection by causing atrophy of lymphatic tissue.

Sex hormones are the *androgens* (male hormones) and *oestrogens* (female hormones) and the effects are similar to those of the hormones produced by the testes and the ovaries. Suprarenal androgens and oestrogens are produced in minute quantities in both sexes and affect the development and function of the reproductive organs and the physical and temperamental characteristics of men and women. Their effects become noticeable when there is over-secretion; for example, a suprarenal tumour in a woman may produce secondary masculine characteristics such as the growth of hair on the face and deepening of the voice.

Under-secretion of the cortical hormones results in a condition known as Addison's disease which produces symptoms of anaemia, muscular weakness, low blood pressure, low blood sugar level, and bronzing of the skin and mucous membranes.

Over-secretion of the cortical hormones results in several dis-orders. Cushing's disease is due mainly to overproduction of the glucocorticoids. There is excess fatty tissue on the trunk and face but not the limbs and sodium is retained so there is oedema, increased plasma volume and a raised pH (alkalosis).

Medulla

The medulla of the suprarenal glands produces two hormones, *adrenaline* and *noradrenaline*, which have effects similar to that obtained by stimulation of the sympathetic nervous system. Because control of the medulla is by way of preganglionic neurones of the sympathetic nervous system, without interruption, response to stimulation is very rapid and creates a set of conditions in which the body is prepared for 'fight or flight'. *Adrenaline* increases the strength and rate of the heart beat; causes dilation of arteries supplying the heart and skeletal muscles but constriction of other arteries; increases rate and depth of respiration; increases blood sugar level by promoting breakdown of liver glycogen and stimu-lates the general metabolic activity of the cells.

The Thymus Gland

The thymus gland is an organ which lies in the lower part of the neck and chest between the lungs over the heart. It varies in size with age and grows until the child is two years of age; it then shrinks so that in the adult only fibrous remnants are found. When fully developed it is greyish-pink in colour and consists of two or three lobes. Its structure resembles that of a lymph node and it is similarly associated with antibody formation. No hormone has yet been isolated from it and it may well belong to the circulatory rather than the endocrine system.

The Pineal Body

The pineal body is a small reddish body about the size of a cherry stone, lying behind the third ventricle of the brain. Its function is unknown. In later life it becomes calcified and acts as a useful land-mark in X-rays of the skull.

21 The Urinary System

The urinary system consists of:

1. The kidneys
2. The ureters
3. The bladder
4. The urethra

The Kidneys

The kidneys are two bean-shaped organs situated in the posterior part of the abdomen, one on each side of the vertebral column, behind the peritoneum. They lie at the level of the twelfth thoracic to the third lumbar vertebrae, though the right kidney is usually slightly lower than the left because of its relationship to the liver.

Each kidney is about 11 cm long, 6 cm wide and 3 cm thick and is embedded in a bed of fat called the *perirenal fat*.

The medial border of each kidney is concave in the centre. This area is called the *hilus* and it is the point at which the blood vessels, nerves and ureters enter or leave the kidney.

The kidney is enclosed in a *capsule* of fibrous tissue which can be easily stripped off. In vertical section the kidney has two distinct parts. The dark outer part is called the *cortex* and the paler inner portion the *medulla*. This leads into the collecting space which is called the *renal pelvis*.

The kidney substance consists of countless minute twisted tubules called *nephrons*. There are over a million in each kidney. Each of the nephrons begins in a cup-shaped expansion called the *glomerular capsule*, from which the tubule leads. Into the cup of each capsule comes a fine branch of the renal artery, forming a tuft of capillaries in close contact with its inner wall; the capillary tuft is called the *glomerulus*. The arteriole bringing blood to the tuft is called the *afferent vessel*, and the arteriole which carries the blood

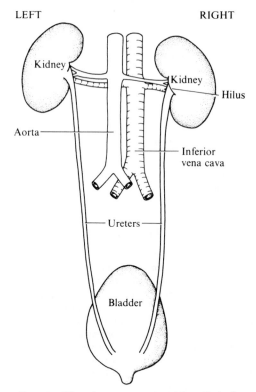

FIG. 160. The urinary system viewed from the back.

away is known as the *efferent vessel*; it is slightly smaller than the afferent vessel. The blood in the tuft is under high pressure because of this and because of its nearness to the abdominal aorta.

The convoluted tubule makes a number of twists, the proximal convolutions, on leaving the capsule and then forms a long loop, the loop of Henle, which dips down into the medulla and passes back to the cortex. The tubule next makes a distal or second series of convolutions, and ultimately empties into a straight collecting tubule in the medulla.

The efferent blood vessel which comes from the capillary tuft or

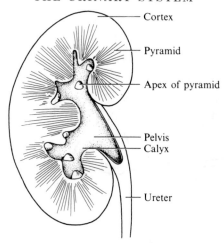

Fig. 161. Diagrammatic section of the kidney.

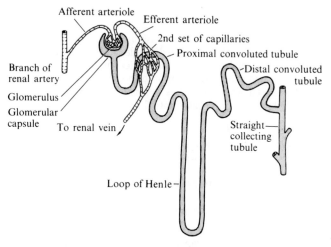

Fig. 162. A nephron and the blood vessels associated with it.

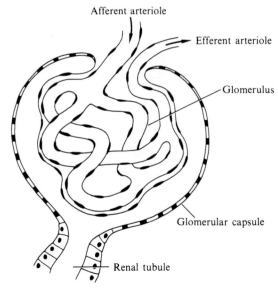

Fig. 163. Diagram of the glomerulus and its capsule.

glomerulus in the capsule divides to form a second set of capillaries, around the walls of the convoluted tubules in the cortex. Thus the blood passes through two sets of capillaries within one organ, which does not happen in any other part of the body. The blood is collected from the second set of capillaries by small veins, which unite with other small veins to empty blood into the renal vein.

The Production of Urine

The function of the kidneys is to secrete and excrete urine. The composition of the blood must not vary beyond certain limits if the tissues are to remain healthy and this regulation depends on the removal of harmful waste products and the conservation of water and electrolytes in the body.

Urine is produced by three processes:

1. *Filtration under pressure* occurs from the glomerulus where only the thin walls of the capillaries and of the glomerular capsule separate the blood from the kidney tubule. The walls of the glomer-

ulus are permeable to water and to other small molecules but they are not permeable to blood cells or to protein. Because the blood in the glomerulus is under pressure some of the constituents pass through into the glomerular capsule. This fluid is known as the *glomerular filtrate* and it has a composition similar to plasma in that it contains glucose, amino acids, fatty acids, salts, urea and uric acid in the same proportions. Blood cells and protein molecules are only filtered if the kidney is diseased. About 600 ml of blood per minute pass through the glomerulus and of this about 125 ml become the glomerular filtrate. If this were all excreted 150 to 180 litres of urine would be passed each day. The average amount of urine passed each day is about 1·5 litres so it is obvious that reabsorption must occur.

2. *Selective reabsorption* occurs because the cells lining the convoluted tubules are able to absorb the water, glucose, salts and their ions which the body needs. In normal health all the glucose is reabsorbed and none is excreted in the urine. Most of the water and salts are also absorbed resulting in the 1·5 litres of fluid which do pass into the collecting tubules containing normally about 2% urea. The acidity of the urine varies somewhat so that the reaction of the blood is maintained at a pH of about 7·4.

3. *Active secretion* occurs because the cells lining the tubule have the ability to secrete some substances from the blood in the second capillary network into the lumen of the tubule.

Reabsorption of water in the distal convoluted tubule is variable and is controlled by the secretion of anti-diuretic hormone from the posterior lobe of the pituitary gland. A decrease in secretion of ADH causes less water to be reabsorbed in the distal tubule, therefore more water is excreted as urine. The reabsorption of salts is controlled by the hormones of the adrenal cortex, especially aldosterone. The production of these hormones is increased or decreased according to the needs of the body to use water or salts and the electrolytes to which they give rise. Nervous control, together with the hormones adrenaline and noradrenaline, maintains the blood pressure at the high level required for filtration in the capillary tufts.

The *medulla* or inner portion of the kidney consists of straight collecting tubules into which the convoluted tubules of the cortex empty. It forms a number of cone-shaped masses which project into the pelvis of the kidney. These are called the *pyramids* of the medulla and are from eight to twelve in number.

The apices of the pyramids project into the pelvis and are covered with the mouths of the fine collecting tubules which pour the urine

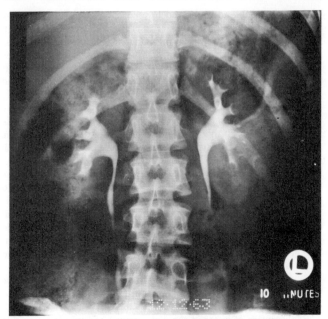

FIG. 164. A pyelogram showing the pelvis of the kidneys filled with an opaque drug which has been excreted following an intravenous injection: note the right kidney is lower than the left. Note also calyces outlining the apices of the pyramids of the medulla.

into the kidney pelvis. The function of the medulla is therefore to collect the urine secreted in the cortex and convey it to the pelvis.

The *pelvis* is an irregular branched cavity, which lies at the root or hilum of the kidney and leads like a funnel to the ureter. Its branches, known as the *calyces* of the pelvis, penetrate into the kidney substance, and each branch receives the apex of one of the pyramids of the medulla. The pyramids pour the urine into the pelvis, which conveys it to the ureters.

The Composition of the Urine

Normal urine, therefore, is formed partly by filtration under pressure from the capsules and partly by reabsorption and by secretion in the tubules. It is an amber-coloured fluid varying in colour according to its quantity. It is acid in reaction and has a specific gravity of 1015 to 1025. (The specific gravity is the weight compared to the weight of an equal quantity of water, water being 1000.)

Urine consists of water, salts and protein waste products, namely urea, uric acid and creatinine. The average composition is: water, 96 per cent; urea, 2 per cent; uric acid and salts, 2 per cent.

The percentage of urea in blood plasma is 0·04 compared to 2 per cent in the urine, hence the concentration has been increased fifty times by the work of the kidney. The salts consist chiefly of sodium chloride and also of phosphates and sulphates, produced partly from the use of the phosphorus and sulphur present in protein foods. These salts must be reabsorbed or got rid of in the quantities necessary to keep the blood at its normal reaction and maintain the water and electrolyte balance. Since this reaction and salt concentration are both essential to the life of the blood corpuscles and the tissue cells this function of the kidney is very important. The normal quantity of urine secreted is 1·5 litres in 24 hours, but it is increased by drinking and by cold weather, and is decreased by reducing fluid intake and by hot weather, exercise, and by fever since these increase sweating. Potassium salts normally filter out and are reabsorbed or excreted as required, to keep the correct level in body fluids. In renal failure, their excretion may be checked so that the amount in the body fluids and tissues rises.

The Ureters

The ureters are the two tubes which carry the urine from the kidneys to the bladder. Each is about 25 to 30 cm in length and is a thick-walled, narrow tube which is continuous with the renal pelvis and which opens into the base of the bladder. It is about 3 mm in diameter but is slightly constricted in three places; (a) at the junction with the renal pelvis; (b) where it crosses the brim of the lesser pelvis; and (c) as it passes through the wall of the bladder. These narrowed portions may be the site of impaction of a ureteric calculus (stone). The ureters, the renal pelvis and the calyces can be seen by radiography following the intravenous injection of a radiopaque substance.

The ureter has an outer fibrous coat continuous with the fibrous capsule of the kidney; a muscular coat which has an outer circular and an inner longitudinal layer, and a lining of mucous membrane which is continuous with the lining of the bladder. The muscular layer of the ureter undergoes peristaltic contractions, usually about four or five times a minute.

The Bladder

The bladder is a reservoir for urine and its size, shape and position vary with the amount of fluid it contains. When empty it lies within the lesser pelvis, but as it becomes distended with urine it expands upwards and forwards into the abdominal cavity.

Both ureters enter and the urethra leaves the bladder at its *base*. An imaginary line drawn to connect these openings outlines the area known as the *trigone*. The *neck* of the bladder is the lowest and most fixed part of the organ; it lies 3 to 4 cm behind the symphysis pubis. The bladder can hold over 500 ml of urine, though this would cause pain; the desire to empty the bladder is normally felt when the organ contains 250 to 300 ml of urine.

The bladder has three coats. The outer serous layer is of peritoneum, but this is found only on the superior surface. The muscular layer contains both circular and longitudinal muscle fibres; there are also two bands of oblique fibres which are situated close to the ureteric openings and which prevent urine flowing back into the ureters. The inner mucous coat is loose and is thrown into rugae when the bladder is empty. The bladder is lined with transitional epithelial tissue which allows for expansion when the organ is full.

The Urethra

The urethra extends from the internal urethral orifice in the bladder to the external urethral orifice.

In the *male* the urethra is 18 to 20 cm in length and serves as a common canal for both reproductive and urinary systems. It is divided into three portions:

1. The *prostatic portion* is about 3 cm in length and is surrounded by the prostate gland. It is lined with transitional epithelium and the *orifices of the prostatic ducts* and the *ejaculatory ducts* open into it.

2. The *membranous portion* is 1 to 2 cm in length and is the narrowest part of the urethra. It passes through the pelvic floor.

3. The *spongy portion* is about 15 cm in length and lies within the penis.

In the *female* the urethra is about 4 cm long and serves the urinary system only. It begins at the internal urethral orifice of the bladder and passes downwards behind the symphysis pubis embedded in the anterior wall of the vagina.

There are internal and external sphincters to the urethra; the internal sphincter is involuntary and the external sphincter is under voluntary control except in early infancy and in nerve injury or disease.

Micturition

Micturition is the passing of urine. Urine is constantly passing into the bladder from the ureters; when there is 200 to 300 ml of urine in the bladder the desire to pass urine occurs due to stimulation of the sensory nerves because of increased tension in the bladder. As the sensory impulses increase in number and frequency the motor impulses cause a reflex contraction of the bladder and relaxation of the internal sphincter. The external sphincter is controlled by the pudendal nerve. When the child has learned to inhibit spinal reflexes micturition can be delayed for a considerable time or can be induced voluntarily.

In disease micturition may be affected in many ways. The sphincter may be paralysed in a state of contraction (spastic paralysis), so that it cannot be relaxed. This will cause *retention* of the urine and the bladder will become overfull. If this is not relieved by catheterization, distension will continue and will distend the orifice which the sphincter guards, allowing a little urine to dribble away continuously though the bladder still remains full. This is called *retention with overflow*. This is bad for both bladder and kidney, causing loss of tone and interference with blood supply to the stretched bladder wall and back pressure on the kidney. It should never be allowed to occur.

The sphincter may be paralysed in a state of relaxation. In such conditions urine will dribble constantly from an empty bladder, as the bladder cannot retain it. This is comparatively rare.

In other cases of nervous disease and in anaesthesia the control of the brain over micturition may be abolished, so that the act again becomes reflex as in the lower animals; the bladder fills and empties reflexly, without the individual being aware of any sensation of fullness or being able to control the relaxation of the sphincter.

22 The Nervous System

The nervous system is the system of communication between the various parts of the body. It is the mechanism by which sensations of all kinds are received from the environment and from the tissues and organs of the body itself; it is responsible for the interpretation of these sensations through dependence on association with similar sensations received in the past and stored in the memory; it is the system by which actions are carried out by the sending of impulses to other parts of the nervous system and other organs of the body.

Nervous Tissue

When nervous tissue is examined under a microscope it is seen to be composed of:

1. Neurones—nerve cells with associated nerve fibres
2. Neuroglia—a special type of connective tissue, found only in the nervous system, which supports the neurones

The *neurone* is the basic tissue of the nervous system. It has a comparatively large *cell body*, though both shape and size vary according to the position of the cell and its function. Each has a clearly defined nucleus and the protoplasm is granular. Nerve cells form the *grey matter* of the brain and spinal cord. The cell has several processes: *dendrites* are short branched processes through which nervous impulses enter the cell and the *axon* (or axis cylinder) is a single fibre through which impulses pass out from the cell. Axons vary considerably in length from a few millimetres to many centimetres and are continuous from cell to termination. Nerve processes, or fibres, form the *white matter* of the brain and spinal cord.

A neurone which has many processes arising from the cell body is

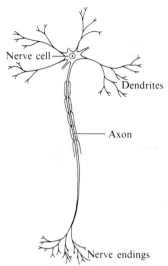

FIG. 165. A typical multipolar neurone.

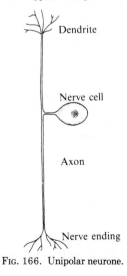

FIG. 166. Unipolar neurone.

termed a *multipolar neurone*. Other types have one process arising from the cell body, which then divides into two branches, one conveying impulses towards the central nervous system, called the axon and one conveying impulses from an organ to the cell; these are *unipolar neurones*. *Bipolar neurones* have two processes, one at each end of the cell; one is a dendrite carrying impulses to the cell and one is an axon conveying impulses from the cell.

Axons, and some dendrites, are surrounded by a thin fatty sheath composed of *myelin* which lies inside the outer covering of connective tissue which is called the *neurilemma* (Fig. 167). The myelin sheath is compressed at intervals and here the neurilemma dips in towards the nerve fibre; these constrictions are called the *nodes of Ranvier*. At these points the nerve fibre has contact with the surrounding tissue fluids, allowing exchange of nutrients and waste materials.

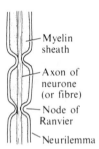

FIG. 167. Myelin sheath.

The myelin sheath is thought to have an insulating effect on the nerve fibre, so that impulses are not transmitted to adjacent nerves or tissues except through the end of the fibre. It also protects the fibre from pressure and injury. Those fibres which have a myelin sheath are called *myelinated* (or *medullated*) *fibres*.

Non-myelinated fibres are found in the autonomic nervous system and in certain parts of the brain and spinal cord.

Nerve cells are quickly damaged by lack of oxygen, toxins and poisonous substances. If they die they cannot be replaced though it may be possible for the function to be taken over by other cells to a limited extent.

A *synapse* is the point of communication between one neurone and the next. The fibrils forming the axon have tiny expanded ends called *end feet* which are close to, but not touching, the dendrites or cell bodies of other neurones. They allow the passage of the nerve impulse in one direction only.

Nerve impulses can also only travel in one direction, into a neurone through the cell body or dendrites and out through the axon. At the synapse there is a short pause to enable a *chemical messenger* to be released to fill the gap between the two neurones and to allow the impulse to pass to the next neurone.

The nervous system is composed of a central nervous system, consisting of the brain and spinal cord, and a peripheral nervous system, consisting of the cranial and spinal nerves and the autonomic nervous system.

Central Nervous System

The Brain

The brain, when fully developed, is a large organ which fills the cranial cavity. Early in its development the brain becomes divided into three parts known as the *forebrain*, the *midbrain* and the *hindbrain*.

The forebrain is the largest part and is called the *cerebrum*; it is divided into right and left hemispheres by a deep *longitudinal fissure*. The separation is complete at the front and the back but in the centre the hemispheres are joined by a broad band of nerve fibres called the *corpus callosum*. The outer layer of the cerebrum is

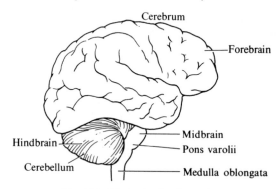

Fig. 168. The brain, showing parts.

Front

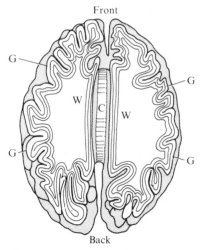

Back

Fig. 169. Section through the cerebrum, viewed from above. G, the grey matter of the convoluted surface; W, the white matter forming the central portion; and C, the corpus callosum, the bridge of white matter joining the two hemispheres of the cerebrum.

called the *cerebral cortex* and is composed of grey matter (cell bodies) thrown into numerous folds or convolutions called *gyri*, separated by fissures called *sulci*. This enables the surface area of the brain, and therefore the number of cell bodies, to be increased greatly. The general pattern of the gyri and sulci is the same in all humans; three main sulci divide each hemisphere into four lobes, each named after the skull bone under which it lies. The *central sulcus* runs downwards and forwards from the top of the hemisphere to a point just above the lateral sulcus; the *lateral sulcus* runs backwards from the lower part of the front of the brain and the *parieto-occipital sulcus* runs downwards and forwards for a short way from the upper posterior part of the hemisphere. The lobes of the hemispheres are the *frontal lobe*, lying in front of the central sulcus and above the lateral sulcus; the *parietal lobe* lying between the central sulcus and the parieto-occipital sulcus and above the line of the lateral sulcus; the *occipital lobe*, which forms the back of the hemisphere and the *temporal lobe* lying below the lateral sulcus and extending back to the occipital lobe.

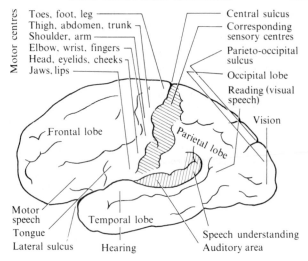

Motor centres
Toes, foot, leg
Thigh, abdomen, trunk
Shoulder, arm
Elbow, wrist, fingers
Head, eyelids, cheeks
Jaws, lips

Central sulcus
Corresponding sensory centres
Parieto-occipital sulcus
Occipital lobe
Reading (visual speech)
Vision

Frontal lobe
Parietal lobe

Motor speech
Tongue
Lateral sulcus
Temporal lobe
Hearing
Speech understanding
Auditory area

Fig. 170. The cerebrum, showing the lobes and the main nerve centres.

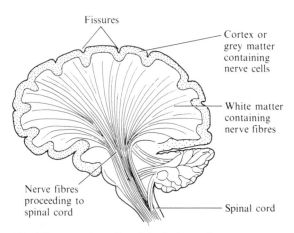

Fissures

Cortex or grey matter containing nerve cells

White matter containing nerve fibres

Nerve fibres proceeding to spinal cord

Spinal cord

Fig. 171. Diagrammatic section of the brain, to show grey matter on the surface and white matter in the centre. The association fibres in the cerebrum are not shown.

The area lying immediately in front of the central sulcus is known as the *pre-central gyrus* and is the motor area from which arise many of the motor fibres of the central nervous system. Immediately behind the central sulcus lies the sensory area, called the *post-central gyrus*, in the cells of which several kinds of sensation are interpreted.

Longitudinal section of a hemisphere shows grey matter (cell bodies) on the outside and white matter (nerve fibres) forming the interior. The nerve fibres connect one part of the brain with other parts and with the spinal cord, but within the white matter groups of nerve cells can be seen forming areas of grey matter. These areas of grey matter are called *cerebral nuclei*. The main function of these areas is coordination of movement and posture of the body: disorders affecting these areas cause jerky movements and unsteadiness.

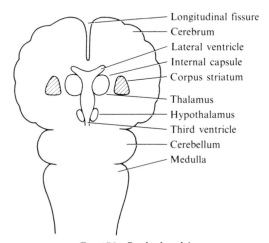

Longitudinal fissure
Cerebrum
Lateral ventricle
Internal capsule
Corpus striatum
Thalamus
Hypothalamus
Third ventricle
Cerebellum
Medulla

Fig. 172. Cerebral nuclei.

The cavities within the brain are called *ventricles*. There are two *lateral ventricles*, a central *third ventricle* and a *fourth ventricle* between the cerebellum and the pons. All are filled with cerebrospinal fluid.

The midbrain lies between the forebrain and the hindbrain. It is about 2 cm in length and consists of two stalk-like bands of white matter called the *cerebral peduncles*, which convey impulses passing to and from the brain and spinal cord, and four small prominences called the *quadrigeminal bodies*, which are concerned with sight and hearing reflexes. The *pineal body* lies between the two upper quadrigeminal bodies.

The hindbrain has three parts:

1. The *pons*, which lies between the midbrain above and the medulla oblongata below. It contains fibres which carry impulses upwards and downwards and some which communicate with the cerebellum.

2. The *medulla oblongata* lies between the pons above and the spinal cord below. It contains the cardiac and respiratory centres which are also known as the vital centres and which control the heart and respiration.

3. The *cerebellum* projects backwards beneath the occipital lobes of the cerebrum. It is connected to the midbrain, the pons and the medulla oblongata by three bands of fibres called the *superior*, *middle* and *inferior cerebellar peduncles* respectively. The cerebellum is responsible for the coordination of muscular activity, control of muscle tone and maintenance of posture. It is continuously receiving sensory impulses concerning the degree of stretch in muscles, the position of joints and information from the cerebral cortex. It sends information to the thalamus and the cerebral cortex.

The midbrain, the pons and the medulla have many functions in common and together are often known as the *brain stem*. This area also contains the nuclei from which originate the cranial nerves.

Spinal Cord

The spinal cord is continuous with the medulla oblongata above and constitutes the central nervous system below the brain. It commences at the foramen magnum and terminates at the level of the first lumbar vertebra; it is about 45 cm in length. At its lower end it tapers off into a conical shape called the *conus medullaris* from the end of which the *filum terminale* descends to the coccyx, surrounded by nerve roots called the *cauda equina*. The cord gives off nerves in pairs throughout its length. It varies somewhat in thickness, swelling out in both the cervical and lumbar regions, where it gives off the large nerve supply to the limbs. The cord is deeply cleft back and front, so that it is almost completely divided into right and left sides like the cerebrum.

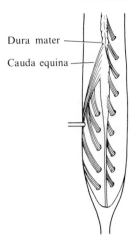

FIG. 173. The cauda equina.

The cord, like the medulla, consists of white matter on the surface and grey matter in the centre. The white matter consists of fibres running between the cord and the brain only, not to the body tissues. It contains:

1. *Motor fibres* running down from the motor centres of the cerebrum and the cerebellum to the motor cells of the cord.

2. *Sensory fibres* running up the cord from the sensory cells of the cord to the sensory centres of the brain.

The *grey matter*, on cutting across the cord, has an H-shaped pattern, with two portions projecting forwards, one on either side, called the *anterior horns*, and two portions projecting backwards, one on either side, called the *posterior horns*.

The *cranial nerves* are twelve pairs of nerves which arise in nuclei in the brain stem. Some are purely sensory, some purely motor and some are mixed nerves carrying both sensory and motor impulses.

The *spinal nerves* are 31 pairs of nerves which arise in the spinal cord. Each has a motor and a sensory component in the anterior and posterior parts of the cord respectively, and the two fibres travel together after they leave the cord.

The *autonomic nervous system* is concerned with the control of internal organs; the function of these organs is not under the control of the will, though they are affected by the emotions.

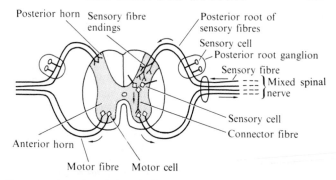

FIG. 174. A section of the spinal cord, showing central grey matter and neurones. Sensory fibres of heat and pressure cross the cord.

The cranial and spinal nerves and the autonomic nervous system, which as mentioned earlier form the peripheral nervous system, are discussed in greater detail later in the chapter.

The Motor System

The motor system is concerned with movement of various parts of the body. As already mentioned (p. 264), the *motor area* is situated in front of the central sulcus in an area called the *pre-central gyrus*. Here the body is represented upside down. At the bottom of the gyrus is a large area for the head and eye; above that is a large area for the hand and arm, then a small area for the trunk and a large area for the leg extending over the top of the cerebral hemisphere (see Fig. 170). There is considerable overlap between these areas. It will be apparent that an area which can undertake a great deal of fine movement such as the hand and arm, will require a larger area of cell bodies than an area such as the trunk which, although greater in extent, does not carry out as much detailed movement. In front of the motor area lies the *pre-motor area* which is concerned with a whole pattern of movement.

Beginning from cells in the motor area the corticospinal fibres pass downwards in a fan-shape (see Fig. 171) and then pass through the *internal capsule*, which lies between the thalamus and the basal ganglia. All the motor fibres serving one side of the body are gathered together in the internal capsule so injury there will cause paralysis in the affected side (hemiplegia). The fibres then pass through the pons to the medulla oblongata where they form a

long narrow projection called a *pyramid*. In the pyramids most of the
motor fibres cross over to the other side, at the *decussation of the
pyramids*, so that fibres that arose in the left cerebral hemisphere
will now be on the right side of the tract and will supply the right
side of the body. The fibres then pass down the spinal cord as the
lateral corticospinal tract. The fibres which did not cross to the
opposite side at the decussation of the pyramids pass down the
spinal cord in the *anterior corticospinal tract* and eventually cross to
the opposite side in the spinal cord.

The fibres pass to the anterior horn of the spinal cord where they
form a synapse with the cell bodies situated there. The fibres then
emerge from the front of the spinal cord as the *anterior root* and join
the corresponding posterior root of sensory fibres to form a *mixed
spinal nerve* (see Fig. 174). As peripheral nerves these end in
branches to various areas, including muscles. The motor fibres
divide into branches and each branch ends in a *motor end plate*
attached to an individual muscle fibre. Sensory fibres have their

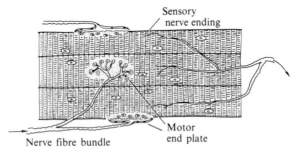

Sensory
nerve ending

Motor
end plate

Nerve fibre bundle

FIG. 175. Muscle, showing motor end plates and sensory nerve endings.

cells in the posterior root ganglion on the spinal nerves and have
endings of various types within the muscles.

The motor area receives impulses from many other parts of the
brain, including the sensory area. From the cortex impulses are sent
to the spinal cord, the motor nuclei in the brain stem, the basal
ganglia, the cerebellum and the pons. Through the various nerve
tracts stimuli are passed through the peripheral nerves to the
skeletal muscles, which are kept in a state of tension called *muscle
tone*. Muscle fatigue is prevented by the use of different sets of

motor fibres successively and the degree of tone depends on the number of fibres being used at any one time.

The term *upper motor neurone* describes the motor fibre within the central nervous system as far as its synapse with an anterior horn cell. A *lower motor neurone* describes an anterior horn cell and its fibre.

The corticospinal tract used to be called the pyramidal tract, so the term *extra-pyramidal system* describes all motor systems other than the corticospinal and corticonuclear tracts. The main function of the system is coordination of muscle movement in the maintenance of body posture so that movements can be performed accurately while still maintaining the desired posture.

The Sensory System

The sensory system is concerned with interpreting the impulses which are constantly stimulating it. Many of these do not reach the level of consciousness because the nervous system deals with them automatically making adjustments to the height of the blood pressure, the rate of the heart beat, the degree of tone in the muscles and many other conditions.

Sensory impulses are transmitted to the central nervous system from the special sense organs, the skin, and from deeper parts of the body.

Special Senses. The impulses of *sight* are conveyed from the retina of the eye through the *optic nerve* (second cranial nerve) to the *optic chiasma* where the medial fibres of each optic nerve cross over to the opposite side. Because of this partial crossing the visual area of the left cerebral hemisphere receives impressions from the outer (temporal) side of the retina of the left eye and from the inner (nasal) side of the retina of the right eye. The impulses are then conveyed to the *visual areas* in the occipital lobes, where they are interpreted. It will be seen that division of the left optic nerve causes blindness of the eye on the same side but that division of the left optic tract causes inability to see the left half of the normal field of vision.

The impulses of *hearing* are conveyed through the vestibulo-cochlear nerve (eighth cranial nerve) to the *pons* where there is a synapse. A second set of fibres carry the impulses to the *medial geniculate body* and a third set to the *auditory areas* in the temporal lobes for interpretation.

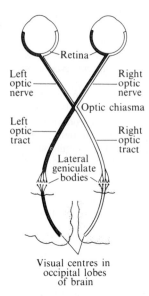

FIG. 176. The visual path from the eyes to the visual centres.

The impulses of *smell* are carried to the *olfactory nerve* (first cranial nerve) and then to the *olfactory bulb* and *olfactory tract* which are under the frontal lobe. Smells are interpreted in various parts of the temporal lobe. There are few *tastes* which can be detected without the sense of smell. Taste impulses are carried by the facial (seventh cranial) and glossopharyngeal (ninth cranial) nerves and are interpreted in the temporal lobe with the corresponding smell.

Sensation from Skin and Muscle. Sensory nerve endings are present in the skin and other tissues. Different kinds of sensation—temperature, touch, pressure and degrees of stretch—require different nerve endings to initiate them. In the skin a sensory nerve fibre may begin as (1) a nerve ending capable of transmitting pain and temperature change, (2) one transmitting light touch, or (3) one transmitting firm pressure. In addition muscles have special structures called *muscle spindles* which

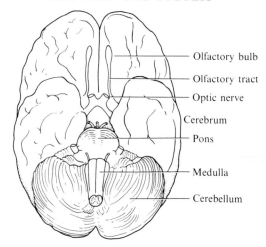

Olfactory bulb

Olfactory tract

Optic nerve

Cerebrum

Pons

Medulla

Cerebellum

FIG. 177. The brain, viewed from below.

respond to the degree of tension to which they are exposed. From all these nerve endings sensory fibres run in the spinal nerves to the posterior nerve roots in the spinal cord. The nerve cell of the first neurone is in the posterior root ganglion. In the head they run in the trigeminal nerve (fifth cranial) and other cranial nerves to the brain.

Nerve fibres carrying light touch and muscle stretch sensations (proprioception) then give off several branches to anterior and posterior horn cells so that each section of the cord becomes a functioning unit. They then pass upwards in the posterior part of the white matter and end in a synapse in the medulla oblongata. A second neurone crosses over to the opposite side and ends at the thalamus. A third neurone carries impulses to the sensory area in the parietal lobe.

Fibres carrying pain and temperature change synapse in the posterior horn. A second neurone crosses over immediately to the opposite side of the cord and ascends to form a synapse in the thalamus. A third neurone passes to the sensory area.

Fibres carrying firm pressure impulses also synapse in the posterior horn. A second neurone crosses over to the opposite side of the cord and ascends, in a different part of the cord, to the thalamus. A third neurone passes to the sensory area.

It can be seen that all sensory fibres eventually cross to the opposite side so that sensation from the left side of the body will be interpreted in the right side of the brain. Also, all sensory neurones form a synapse in the thalamus.

The *thalamus* (see Fig. 172) is responsible for sorting out the mass of sensory information being fed into the body and passing it on to the cerebral cortex, when necessary, or to other areas of the brain as appropriate. The *hypothalamus* is concerned with the stability of the internal organs of the body. It controls water balance, regulates appetite, temperature and sleep and plays a part in the control of emotion.

The sensory area of the cortex lies in the parietal lobe behind the central sulcus. Like the motor area the body is represented upside down, with large areas for the face, head and hand at the lower end and the smaller areas for the arm, trunk and leg which are comparatively less sensitive.

The Meninges

The meninges are the protective membranes covering the central nervous system. There are three layers.

The outer layer is called the *dura mater*. It is a tough fibrous membrane which has two layers, the outer layer lining the inner surface of the skull and forming the periosteum. At the foramen magnum this layer continues as the periosteum on the outer surface of the skull. The inner layer of dura projects inwards in certain places to form a double layer which separates parts of the brain and helps to maintain them in position. The *falx cerebri* is one such fold, between the two cerebral hemispheres; another is the *tentorium cerebelli* between the cerebrum and the cerebellum. The two layers

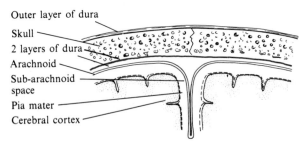

Outer layer of dura
Skull
2 layers of dura
Arachnoid
Sub-arachnoid space
Pia mater
Cerebral cortex

FIG. 178. Diagram of the meninges.

of the dura are in contact with one another for the most part, but are separate when they enclose a venous sinus. The inner layer of dura mater also encloses the spinal cord and continues down as far as the sacrum.

The *sub-dural space* is a potential rather than an actual space which lies between the dura mater and the arachnoid mater.

The *arachnoid mater* is a delicate membrane which lies immediately beneath the dura and dips down with it between the main portions of the brain.

The *sub-arachnoid space* lies between the arachnoid mater and the pia mater and is filled with cerebrospinal fluid. Between the cerebellum and the medulla oblongata there is a comparatively large space called the *cisterna magna*. This is used to obtain a sample of cerebrospinal fluid in small children. The arachnoid mater accompanies the dura as a covering to the spinal cord and extends to the sacrum.

The *pia mater* is a thin, vascular membrane which is in contact with the surface of the brain and spinal cord and dips into all the convolutions.

The Cerebrospinal Fluid. The cerebrospinal fluid is a clear, colourless liquid which fills the sub-arachnoid space and the ventricles of the brain. It is secreted by the *choroid plexuses* in the ventricles, and passes from the two lateral ventricles, which communicate with each other and with the third ventricle through the interventricular foramen, to the third ventricle and then through a narrow tube, called the aqueduct, into the fourth ventricle. There are three small openings in the roof of the fourth ventricle through which the cerebrospinal fluid passes into the sub-arachnoid space in which it circulates around the outside of the brain and spinal cord. It is finally absorbed through the *arachnoid granulations*, which are small projections of arachnoid mater, into the venous sinuses.

The cerebrospinal fluid is similar to blood plasma in composition although it has only a small amount of protein. There are about 120 ml of fluid altogether with a pressure of 60 to 150 mm of water. It usually contains 200–300 mg protein per litre and about 2.8–4.4 mmol glucose per litre. These amounts may be altered in disease.

The main function of the cerebrospinal fluid is to protect the brain and spinal cord by forming a water cushion between the delicate nerve tissues and the walls of the bony cavities in which they lie. It enables the pressure within the skull to be kept constant and it carries away waste and toxic substances.

Peripheral Nervous System

The Cranial Nerves

The cranial nerves originate or terminate within the brain. Some are motor nerves, some are sensory nerves, and some are mixed nerves (see Table 16).

The Spinal Nerves

The spinal nerves are divided into groups according to the region of the cord from which they arise. There are:

1. Eight pairs of *cervical* nerves, one above the atlas and one below each of the cervical vertebrae
2. Twelve pairs of *thoracic* nerves, one below each thoracic vertebra
3. Five pairs of *lumbar* nerves ⎫ derived from the cauda
4. Five pairs of *sacral* nerves ⎬ equina
5. One pair of *coccygeal* nerves ⎭

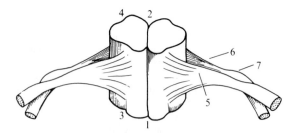

FIG. 179. A section of the spinal cord, showing a pair of spinal nerves arising from it. 1 and 2, fissures which divide the right and left sides of the cord; 3 and 4, smaller fissures at the side from which the nerves arise; 5, the anterior motor root of the spinal nerve; 6, the posterior or sensory root, on which the posterior root ganglion (7) shows as a swelling close to the point where the two roots join.

The spinal nerves give off short posterior branches which supply the muscles of the back of the neck and trunk, and long anterior branches, which provide the nerves of the limbs and the sides and front of the trunk.

In certain regions these nerves branch immediately they leave the spinal canal, and the branches join up with one another to form the

Table 16
Cranial Nerves

Name	Type	Function and Distribution
1. Olfactory nerve	Sensory	The nerve of smell. Starts in the nose and passes to the olfactory bulb
2. Optic nerve	Sensory	The nerve of sight. Starts in the retina and passes to the lateral geniculate body
3. Oculomotor nerve	Motor	Arises in the midbrain and ends in the muscles which move the eye
4. Trochlear nerve	Motor	As the third cranial nerve
5. Trigeminal nerve	Motor and sensory	Supplies the muscles of mastication and has three sensory branches— ophthalmic, maxillary and mandibular
6. Abducent nerve	Motor	Arises in the pons and ends in one of the muscles moving the eye
7. Facial nerve	Motor and sensory	Supplies the muscles of facial expression and is sensory from the tongue
8. Vestibulocochlear nerve	Sensory	Branches from the ear and the cochlea give hearing and the sense of balance
9. Glossopharyngeal nerve	Motor and sensory	The nerve of taste. Also sends motor fibres to the pharynx
10. Vagus nerve	Motor and sensory	Supplies the digestive tract controlling both secretion and movement
11. Accessory nerve	Motor	Supplies the muscles of the neck, most of the pharynx and soft palate
12. Hypoglossal	Motor	Supplies the tongue

nerves supplying the various muscles and parts. This inter-branching is called a plexus. Plexuses are formed in all regions except the thoracic region.

The *cervical nerves* form two plexuses:

1. The *cervical plexus*, which supplies muscles of the neck and shoulder, and also gives off the *phrenic nerve* supplying the diaphragm.

2. The *brachial plexus*, which supplies the upper limb.

The *brachial plexus* gives off three main nerves—the radial, median and ulnar nerves. The *radial nerve* runs round the back of the humerus and down the outer side of the forearm. It supplies the extensor muscles of the elbow, wrist and hand. It is exposed to pressure in the armpit and against the humerus, and injury to it is the cause of wrist-drop, when the joint is flexed and cannot be extended. This may result from the use of badly padded, cheap crutches, with no hand rest, pressing in the armpit or from pressure by the edge of the operation table on the nerve against the humerus, if the patient's arms are allowed to hang down during an operation.

The *ulnar* and *median nerves* run down the inner side and middle

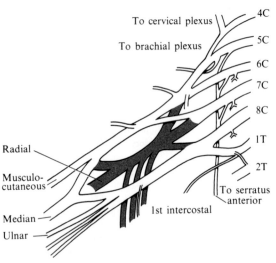

FIG. 180. The brachial plexus and the main nerves arising from it.

of the limb respectively, and supply the flexor muscles of the wrist and hand. Injury to them causes hyperextension and gives rise to the 'claw-like' hand, the unopposed extensor muscles coming into play. A fourth smaller nerve, the *musculocutaneous nerve*, supplies the flexors of the elbow-joint, the biceps, and brachialis muscles. It is the ulnar nerve, crossing in the groove between the back surface of the internal epicondyle of the humerus and the olecranon, which is knocked when we say we have 'knocked our funny-bone', the tingling pain passing down the nerve to the hand.

The *thoracic nerves* supply the muscles of the chest and the main part of the abdominal wall.

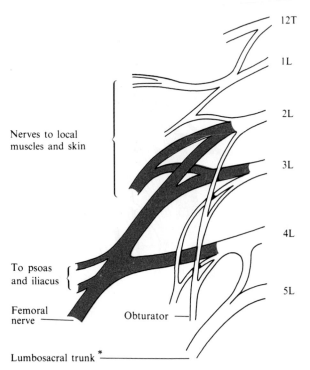

FIG. 181. The lumbar plexus and the main nerves arising from it. *This trunk joins the sacral plexus.

The *lumbar nerves* form the lumbar plexus, which gives off one main nerve, the *femoral nerve*. This runs down beside the psoas muscle under the inguinal ligament into the front of the thigh and supplies the muscles there. The lumbar plexus also gives off branches to the lower plexus wall.

The *sacral nerves* with branches from the fourth and fifth lumbar nerves form the sacral plexus, which gives off one large nerve, the *sciatic nerve*. This is the largest nerve in the body. It leaves the

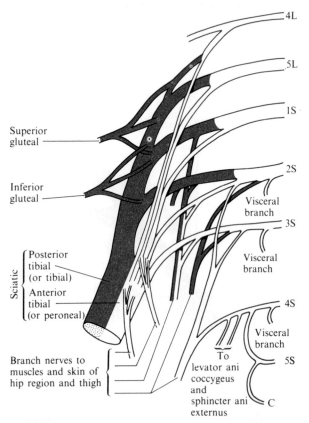

FIG. 182. The sacral plexus, showing the main nerves arising from it.

pelvis by the sciatic notch, runs across the back of the hip-joint and down the back of the thigh, supplying the muscles there. It divides above the knee into two main branches:

1. The *peroneal nerve*, which supplies the muscles of the front of the leg and foot.
2. The *tibial nerve* which supplies the muscles of the back of the leg.

The sciatic nerve therefore supplies the whole of the leg below the knee except for a small sensory branch from the femoral nerve.

The *coccygeal nerves*, with branches from the lower sacral nerves, form a second small plexus on the back of the pelvic cavity, supplying the muscles and skin in that area, e.g. the muscles of the perineal body, the external sphincter of the anus, the skin, and other tissues of the external genitals and the perineum, etc.

Both the sacral and the coccygeal nerves also give off branches to the sympathetic ganglia in the pelvic area.

The Autonomic Nervous System

The autonomic nervous system supplies nerves to all the internal organs of the body and the blood vessels. It is so named because these organs are self-controlled (auto = self) and not under the control of the will. The functioning of the internal organs normally takes place without any conscious knowledge. The will does not normally affect them, but the emotions do. They are affected by the hypothalamus.

The autonomic nervous system is for the most part efferent. It is made up largely of efferent neurones, both motor, supplying the involuntary muscles of the walls of organs such as the stomach, intestines, bladder, heart and blood vessels, and secretory, supplying the glands such as the liver, pancreas and the kidney. There are some afferent fibres, but they are comparatively few in number, as the internal organs are almost insensitive. As a result disease may attack and destroy them without causing pain, and the pain which does occur is due to a large extent to inflammation of the lining membrane of the cavity in which they lie; for example, tuberculosis or pneumonia may affect the lung tissue without any pain, but as soon as the pleura is involved sharp pain is felt. In the same way it hurts to cut through the abdominal wall, but when a piece of bowel has been brought out of the abdomen it can be cut without

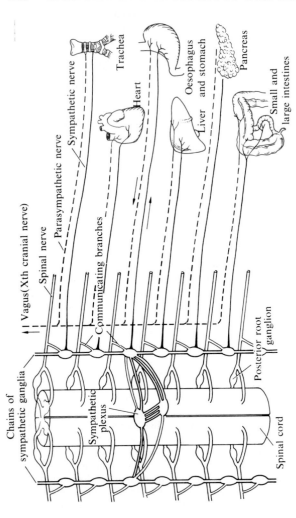

Fig. 183. Diagrammatic representation of the parts of the autonomic nervous system. The sympathetic nervous system is shown by a continuous line, and the parasympathetic nervous system by broken lines.

causing the patient any pain, as the nurse may see in cases of colostomy.

The autonomic nervous system consists of two parts:

1. The sympathetic nervous system
2. The parasympathetic nervous system

The *sympathetic* nervous system consists of a *double chain of ganglia* running down the trunk just in front of the vertebral column in the cervical, thoracic and lumbar regions. These ganglia are linked to one another by nerves. They receive nerves from the thoracic and upper lumbar regions of the spinal cord. They give off

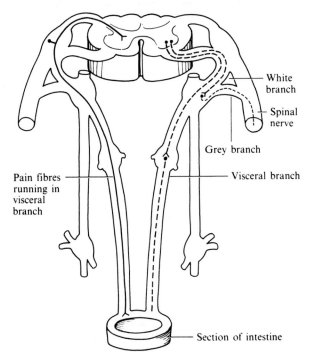

White branch

Spinal nerve

Grey branch

Visceral branch

Pain fibres running in visceral branch

Section of intestine

FIG. 184. Section of the spinal cord showing efferent and afferent pathways of the sympathetic nerves and the link with the spinal nerve. (After Cockett.)

nerves supplying the internal organs, the *visceral branches*, and nerves running back to the spinal nerves, the *parietal branches*. These nerves supply the blood vessels, the sweat and sebaceous glands and the muscles which raise the hairs of the skin making them 'stand on end'. In certain regions where there are many organs requiring a nerve supply there are additional ganglia between the two chains, linked up by nerves with the chains and with one another, giving off nerves to the neighbouring organs; these are called plexuses, e.g. the cardiac plexus lies behind the heart in the thoracic region, and the solar plexus lies just below the diaphragm, where stomach, liver, kidneys, spleen and pancreas are all found.

The *parasympathetic* nervous system consists chiefly of the vagus nerve, which gives off branches to all the organs of the thorax and abdomen but does also include branches from other cranial nerves (the third, seventh and ninth), and nerves from ganglia in the sacral region of the vertebral column.

The Functions of the Autonomic System. All the internal organs have, therefore, a double nerve supply, sympathetic and parasympathetic, and the two sets of nerves have in each case opposite actions, one *stimulating* and the other *checking* the activity of the organ. The arrangement is similar to a motorcar, with an accelerator pedal to make it move faster and a brake pedal to check it.

The *sympathetic* nerves have a stimulating and quickening effect on the heart and on the respiratory system, but a checking effect on digestion. They improve the circulation and cause dilation of the bronchial tubes, increasing the air intake, but they stop the secretion of digestive juices from the salivary glands, and throughout the alimentary canal, and check peristaltic action in its wall. These nerves are stimulated by strong emotion, such as fear, anger and excitement. (It is because of this effect of the emotions that they are called sympathetic.) Their functions are thus closely linked to those of the adrenal medulla, which they stimulate. They help to enable the body to respond to emotion, since they provide the muscles with a better supply of blood which is rich in oxygen. This enables the individual to run away when frightened or to fight when angry, the instinctive response to these emotions. On the other hand, they are responsible for the arrest of the digestion of food in strong emotion, and thus may produce vomiting and emptying of the bowel, the organs getting rid of the contents which they cannot cope with for the time being.

The *parasympathetic* nerves have exactly opposite effects, stimulating the digestive system and producing both a copious flow of digestive juices and peristaltic action. On the other hand, the vagus slows the heart, reducing the circulation, and has a checking effect on the respiratory system, contracting the bronchial tubes. These nerves are stimulated by pleasant emotions. As a result, happiness and a contented mind tend to improve digestion. Pavlov was able to show this in a dog with a gastric fistula. When the dog was shown a bone which pleased him, gastric juice began to pour into the stomach by reflex action through the vagus nerve. The bringing of a cat into the room angered the dog and the flow was checked, the sympathetic nerves being stimulated. Pavlov also noticed that if a bell was rung before the dog's food was brought in each day, after a time the ringing of the bell without bringing the food would cause a flow of gastric juice. This is termed a *conditioned reflex*. The animal had learned to associate the two things, so that either would produce the same reflex response. It is the same with human beings. The ringing of a dinner bell will cause secretion of digestive juices, just as will the sensation of the smell and sight of pleasant, well-served food. In the invalid, whose digestion is impaired, it is therefore of special importance to serve foods appetizingly, and choose dishes which will give the patient as much pleasure as possible.

Reflex Actions

Reflex action is the result of the stimulation of the motor cells by stimuli brought in by afferent neurones from the tissues. Incoming stimuli can therefore, in addition to causing sensation, give rise to action. They only produce sensation if they are passed on to the sensory centres of the brain. On the other hand, in cord and brain they may stimulate the motor cells and give rise to action—*reflex action*. Sensory stimuli are pouring into the cord and brain from the tissues all the time. If they reach the sensory centres of the cortex of the cerebrum and stimulate them they produce sensations of which we are conscious. If they stimulate motor cells they produce reflex action, e.g. the touch of something hot on the skin causes immediate withdrawal of the part, a tap on the patella ligament causes contraction of the quadriceps extensor muscle and produces the 'knee-jerk', stroking of the sole of the foot causes the toes to be drawn down. In the case of the first action it may be claimed that because the heat causes an unpleasant sensation we want to withdraw the part and the action is voluntary. On the other hand, the action occurs in an animal when the spinal cord has been severed in the neck, so that no stimuli can reach the brain, and is, in fact, more marked than in an uninjured animal. The sensory stimulus is brought into the cord by sensory fibres, transmitted by connector fibres to the motor cells of the anterior horn, and passed out by the motor fibres to the muscles. In the case of the knee-jerk the sensory stimulus is carried into the lumbar region of the cord by afferent fibres and stimulates the motor cells, which control the quadriceps.

The reason for the reflex being more marked when the cord is cut off from the brain is that the cerebral centres have an *inhibiting effect* on reflex action. Reflex action is not produced by the will; it is a mere response to environment, and is the only type of action found in the lower forms of life. With the development of the cerebrum, control of reflex action and the partial replacement of reflex action by voluntary action takes place. If we touch something hot the natural reflex is to withdraw, as the inexperienced puppy does when it touches a hot cinder in the grate. On the other hand, if we pick up a hot plate with our dinner on it, appreciating the value of the plate

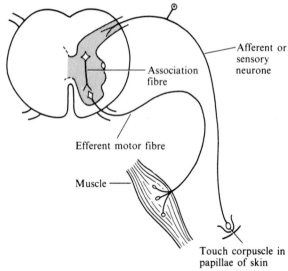

Afferent or
sensory
neurone

Association
fibre

Efferent motor fibre

Muscle

Touch corpuscle in
papillae of skin

FIG. 185. A reflex arc.

and the dinner, we can continue to hold the plate and put it safely down at the expense of burnt fingers. The reflex is inhibited. If we have been warned and expect the plate to be hot, this control is more easily established. In the case of the pure reflex, where the action which is produced has no value, as in the knee-jerk and the drawing down of the toes on tickling the sole of the foot, there is still some inhibiting effect from the brain, and if the nerve path from the brain to muscle is destroyed by injury or disease above the anterior horn cell the action is much more marked.

Actions which are in the first place voluntary become in a sense reflex, e.g. standing is in the first place a voluntary act which is carried out by the exercise of the will. When we have learnt to keep our balance on two feet we learn to do it by the sensations from the skin of our feet, and from muscles and joints, and the sensory organs of balance, and we can stand without voluntary effort unless disease affects the sensory nerves. In the same way when we learn to knit it is at first a voluntary action. Gradually our fingers and arms learn the feel of the wool, needles and actions, and we can knit without attention unless we lose the sensation in the parts or are learning a new complicated pattern.

Reflex action may occur at three different levels in the nervous system:

1. The spinal reflex, e.g. the knee-jerk
2. Reflexes occurring at the base of the brain, e.g. sneezing, coughing, vomiting, walking (cerebellar)
3. Reflexes occurring in the cerebrum, and involving use of the association fibres of the brain

23 The Ear

The ear is the organ of hearing and also plays an important part in the maintenance of balance. The external ear, the middle ear and the cochlea of the internal ear are concerned with hearing; the semicircular canals, the utricle and the saccule of the internal ear are concerned with balance.

The Structure of the Ear

The *external ear* has two parts; the auricle and the external acoustic meatus. The *auricle* projects from the side of the head. It is composed of a thin piece of elastic fibrocartilage, covered with skin,

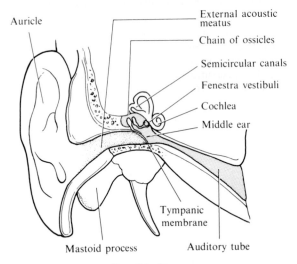

Auricle

External acoustic meatus

Chain of ossicles

Semicircular canals

Fenestra vestibuli

Cochlea

Middle ear

Tympanic membrane

Mastoid process

Auditory tube

Fig. 186. The ear.

which funnels sound waves towards the external acoustic meatus. The *external accoustic meatus* is a tubular passage about 4 cm long leading into the temporal bone. The outer one-third has walls of cartilage and the inner two-thirds walls of bone; the canal is curved, running first forwards and upwards, then backwards and upwards and finally forwards and slightly downwards. These curves may be straightened by gentle traction on the auricle; in adults the pull is upwards and backwards, in children backwards only and in infants downwards and backwards. The inner end of the meatus is closed by the *tympanic membrane*. The skin lining the cartilaginous meatus contains hair follicles and numerous glands which secrete *cerumen*. These protect the canal from foreign bodies by entangling dust and other particles, but the cerumen may itself block the canal if it accumulates, and will then require removal by syringing.

The *middle ear* is a small space within the temporal bone. The tympanic membrane separates it from the external ear and its further (medial) wall is formed by the lateral wall of the internal ear. The cavity is lined with mucous membrane and is filled with air which enters from the pharynx through the auditory tube. This equalizes air pressure on both sides of the tympanic membrane. It contains a chain of three tiny bones, called *ossicles*, which transmit the vibrations of the tympanic membrane across to the internal ear. The tympanic membrane is thin and semi-transparent and the handle of the *malleus*, the first of the ossicles, is firmly attached to the inner surface. The *incus* articulates with the malleus and with the *stapes*, the base of which is attached to the fenestra vestibuli, which leads to the internal ear. The posterior wall of the middle ear has an irregular opening which leads into the *mastoid antrum* and this in turn leads to a number of mastoid air cells. These are air-filled cavities within the bone which, like the nasal sinuses, may become infected.

The *internal ear* lies in the petrous part of the temporal bone. It consists of two parts, the bony labyrinth and the membranous labyrinth.

The *bony labyrinth* is again divided into three parts; the vestibule, the cochlea and the semicircular canals.

The *vestibule* adjoins the middle ear through two openings; the *fenestra vestibuli*, which is filled by the base of the stapes, and the *fenestra cochlea* which is filled by fibrous tissue. At the back are openings into the semicircular canals and at the front an opening into the cochlea.

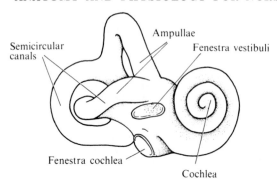

Fig. 187. The bony labyrinth.

The *cochlea* is concerned with hearing. It is a spiral tube which makes two and three-quarter turns round a central pillar of bone called the *modiolus*. The tube is divided lengthwise into three separate tunnels by two membranes, the basilar membrane and the vestibular membrane, which stretch from the modiolus to the outer wall. The outer tunnels are the scala vestibuli above and the scala tympani below. These tunnels are filled with perilymph and they join at the top of the modiolus. The lower end of the scala tympani is closed by the fibrous fenestra cochlea. The middle tunnel is called the cochlear duct and is filled with endolymph. It is the same shape as the bony labyrinth and is called the membranous labyrinth. With-

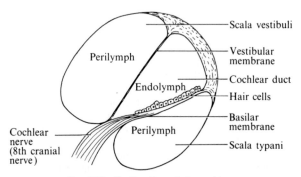

Fig. 188. Section through the cochlea.

in the cochlear duct are the special nerve endings of the auditory nerve called hair cells.

The *semicircular canals* are three in number and they lie above and behind the vestibule in three different planes of space, one vertical, one horizontal and one transverse. They contain perilymph. When the position of the head is altered the movement of the endolymph stimulates special cells with hair-like processes situated at the end of each canal. This information helps in the maintenance of posture, though in daylight it is mainly the responsibility of the eyes to supply information about the position of the head in space. Over-stimulation of the fluid in the semicircular canals causes giddiness.

The *membranous labyrinth* is contained within the bony labyrinth although it is much smaller. It includes the utricle, the saccule, the semicircular ducts and the cochlear duct.

The *utricle* and *saccule* are two small sacs in the vestibule which communicate with each other through a connecting tube. They contain patches of sensory hair cells which are stimulated by the action of gravity on small crystals (otoliths) which adhere to them.

The *semicircular ducts* are similar in shape to the semicircular canals and lie within them but are only a quarter of the diameter. They contain endolymph.

The *cochlear duct* is a spiral tube within the bony canal of the cochlea and lying along its outer wall. Its roof is formed by the vestibular membrane and its floor by the basilar membrane and its outer wall by the bony wall of the cochlea.

The Mechanism of Hearing

The sound wave is a wave of *vibration* of the air set up by the vibration of an object. The vibration of a fiddle-string or the vocal cord, for instance, sets up vibration of the air in contact with it and produces waves of vibration which spread out in all directions, like the ripples set up in a still pond when a pebble is thrown into it.

To produce sound, vibrations must be within certain rates. The human ear is stimulated only by vibrations at rates beween 30 and 30 000 per second. The slow vibrations produce low notes and the quick vibrations produce high ones. It is for this reason that a man's voice is lower than a woman's, for his vocal cords are longer and vibrate more slowly while the woman's are shorter and vibrate more quickly. It is the rapid growth in the larynx and the consequent lengthening of the boy's vocal cords at puberty which causes the breaking of the voice.

Sound waves travel at a speed of 340 m per second. They travel much more slowly than light rays, hence the flash of lightning is seen before thunder is heard, and the farther away the storm the longer the interval between the two.

Sound waves normally are carried by air, but they also pass through solid bodies; in fact, a solid carries sound more readily than air. Thus, putting an ear to the ground, it is possible to hear footsteps at a greater distance than through the air. It is, however, normally air with which the ear is in contact.

Hearing is due to the sound waves making the tympanic membrane vibrate, which sets the ossicles and fenestra vestibuli vibrating, which in turn causes vibration of the perilymph. As fluid is incompressible, the perilymph can vibrate only if the fenestra cochlea is able to bulge outwards as the fenestra vestibuli bulges inwards. Hence the need for two windows in the internal ear. Vibration of the perilymph gives rise to vibration of the endolymph,

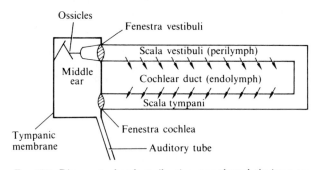

Fig. 189. Diagram to show how vibrations pass through the inner ear.

and this affects the little hairs which jut into it and stimulates the endings of the vestibulo-cochlear nerve in the membranous cochlea. The nerve carries the stimulus to the centre of hearing in the temporal lobe of the brain, where it is appreciated and interpreted.

The appreciation of sound will result from any stimulus brought by the auditory nerve to the centre of hearing, but the meaning given to the sound will depend on previous experience and the power of reasoning.

24 The Eye

The eye is the organ of sight and it is situated in the orbit which protects it from injury.

The Structure of the Eye

The eye is spherical in shape and is embedded in fat. It has three coats: an outer fibrous coat, a vascular, pigmented coat and an inner nervous coat.

The outer *fibrous coat* has two parts. The posterior part is opaque and is called the *sclera*; it is a firm membrane which preserves the shape of the eyeball. Its external surface is white and forms the white of the eye. The anterior part of the *sclera* is covered with conjunctiva, which is reflected on to it from the inner side of the

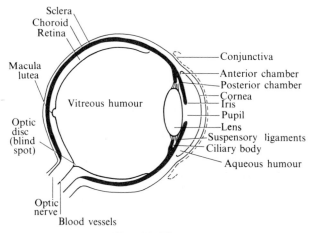

FIG. 190. The eye.

eyelid and is continuous with the corneal epithelium covering the cornea. The *cornea* is the anterior part of the fibrous coat. It projects a little from the surface of the eye and is transparent, allowing light rays to enter the eye and bending them to focus on the retina (refraction).

The *vascular, pigmented coat* has three parts. The *choroid* lines all but the front part of the eye. It is dark brown in colour and supplies blood to the other layers of the eye, particularly the retina. The *ciliary body* is a thickened part of the middle coat containing muscular and glandular tissue. The *ciliary muscles* control the shape of the lens, enabling it to focus light rays from near or far away as required. They are known as the muscles of *accommodation*. The *ciliary glands* produce a watery fluid, the *aqueous humour*, which fills the eye in front of the lens and passes into the veins through small openings in the angle between the iris and the cornea. The *iris* is the coloured part of the eye. It lies between the cornea and the lens and divides the space between them into anterior and posterior chambers. The iris contains muscular tissue arranged in circular and in radiating fibres; the circular fibres contract the pupil and the

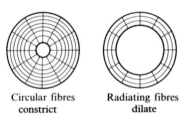

Circular fibres   Radiating fibres
constrict   dilate

FIG. 191. The pupil.

radiating fibres dilate it. There is a circular opening in the centre called the *pupil* which is contracted in bright light to prevent too much light entering the eye and dilated in poor light to allow as much light as possible to reach the retina.

The inner lining of the eye is called the *retina*. It is a delicate membrane adapted for the reception of light rays and contains many nerve cells and fibres. It is made of *rods* and *cones* which are thought to have separate functions. The cones are more numerous in the centre of the eye and are responsible for detailed vision and colour perception; the rods are more numerous around the outer edge of the

Fig. 192. Layer of rods and cones on the outside of the retina.

retina and are sensitive to the movements of objects within the field of vision. They contain a pigment called *visual purple* for the synthesis of which vitamin A is required; lack of vitamin A in the diet may result in *night blindness*. Near the centre of the back of the retina there is an oval yellowish area called the *macula lutea* where only cones are present and which is the area where vision is most perfect. About 3 mm to the nasal side of the macula lutea the optic nerve leaves the eye; this area is called the *optic disc* and because it is insensitive to light it is called the blind spot. An object can only be opposite the blind spot in one eye at a time. To mark a blind spot, mark a piece of paper as shown below:

Now shut the left eye, fix the right eye steadily on the cross, and move the paper slowly backwards and forwards on eye level. At a certain point the dot will disappear because it is opposite the blind spot.

The eye contains:

1. The aqueous humour
2. The vitreous humour
3. The lens

The *aqueous humour* has already been described on page 292.

The *vitreous humour* is a colourless, transparent jelly-like substance which maintains the shape of the eyeball.

The *lens* is situated immediately behind the iris. It is a transparent, biconvex body enclosed in a transparent, elastic capsule from which ligaments pass to the ciliary body. These *suspensory ligaments* hold the lens in position and are the means by which the ciliary muscles exert pull on the lens, altering its shape for near or far vision.

The Mechanism of Sight

As light rays pass through the transparent cornea, aqueous humour and lens they are bent, a process known as *refraction*. This makes it possible for light from a large area to be focused on a small area of the retina. Parallel light rays striking a convex lens are bent towards a focal point on the retina. If the object is less than seven

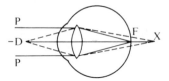

FIG. 193. The normal eye at rest, P = Parallel rays from a distant object focused by the lens on the retina (F). Dotted lines show the diverging rays from a near object (D) from which the focus falls behind the retina at X.

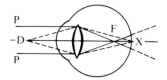

FIG. 194. The normal eye during accommodation. Dotted lines from a near object (D) are focused on the retina at point X by the more curved lens. P = Parallel rays from a distant object which are now focused in front of the retina (F).

metres away the curvature of the lens must be increased to enable the focus to be on the retina. This is called *accommodation*. Far vision can be achieved with the lens in its normal resting position.

Some people are naturally *short-sighted*. This is due to the fact that their eyes are too long, so that the retina is farther from the lens than it should be and the focusing point lies in front of it. In this case the near object can be seen with the flatter lens of the eye at rest

FIG. 195. A convex lens, showing how parallel light rays are brought to focus at F.

(normally used for far sight), but for far sight concave glasses are necessary to throw the focus point farther back.

Others are naturally *long-sighted*. This is due to the fact that their eyes are too short, so that the retina is too close to the lens and the focusing point lies behind it. In this case the far object can be seen by using the thicker, more curved lens used by the normal person for near sight, as this will bend the light rays more and bring the focusing point forward; for near sight, since they are already using the more curved lens, they must be provided with convex glasses to bend the light rays from the near object still more acutely.

The eye is moved within the orbit by six orbital muscles which are ribbon-like and are attached to the sclera. These muscles pull on the eyes and coordinate their movement so that both eyes focus on the same object. A weakness in one or more muscles will cause one eye to deviate, a condition that is commonly known as a *squint*.

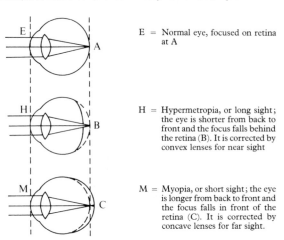

E = Normal eye, focused on retina at A

H = Hypermetropia, or long sight; the eye is shorter from back to front and the focus falls behind the retina (B). It is corrected by convex lenses for near sight

M = Myopia, or short sight; the eye is longer from back to front and the focus falls in front of the retina (C). It is corrected by concave lenses for far sight.

FIG. 196. The normal eye, hypermetropia and myopia.

Protection of the Eyes

The eyes are very delicate organs and are protected by the eye-brows, the eyelids and the lacrimal apparatus, as well as by the bony orbits in which they lie embedded in fatty tissue.

The overhanging *brows* protect them from injury and excessive light, while the hairs entangle sweat and prevent it from running into the eyes.

The *eyelids* consist of a plate of fibrous tissue covered by skin and lined with mucous membrane. The edges of the lids are provided with hairs, the eyelashes, which keep out dust, insects and too much light. The transparent mucous membrane which lines the lids is reflected over the front of the eyeball and is called the *conjunctiva*. This results in the formation of upper and lower conjunctival sacs under the upper and lower lids respectively. Dust and bacteria tend to stick to the moist surface of this membrane, and to keep it clean it is constantly washed by the lacrimal apparatus.

The *lacrimal apparatus* consists of:

1. The *lacrimal gland*, lying over the eye at the outer side and secreting lacrimal fluid into the conjunctival sac.

2. Two fine canals, called *lacrimal canaliculi*, leading from the inner angle of the lids to the lacrimal sac.

3. The *lacrimal sac*, which lies at the inner angle of the eyelids in the groove on the lacrimal bone.

4. The *naso-lacrimal duct*, which runs from the lacrimal sac down to the nose.

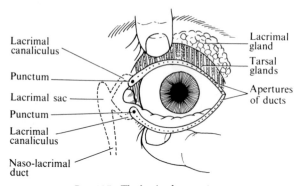

FIG. 197. The lacrimal apparatus.

The opening into the canaliculi can be seen at the inner angle of the eyelids and is called the *punctum*.

The fluid secreted by the lacrimal glands washes over the eyeball and is swept up by the blinking action of the eyelids. The muscles which cause the blink press on the lacrimal sac and contract it so that as they relax the sac expands and sucks the fluid from the edges of the lids along the fine canals into the sac; from this it runs by gravity down into the nose. Thus the window which admits light into the eye is constantly irrigated by a gentle stream of fluid, which keeps it clean and washes away germs and harmful substances. The fluid is composed of water, salts and an antibacterial substance called *lysozyme*.

25 The Skin

The skin covers the body and protects the deeper tissues. It contains the endings of many sensory nerves and is important in the regulation of body temperatures.

Structure of the Skin

The skin has two layers:

1. The epidermis, or outer layer
2. The corium

The *epidermis* is non-vascular and consists of stratified epithelium. It is very thick, hard and horny on such areas as the palms of the hands and the soles of the feet and is much thinner and softer over other parts, such as the trunk and the inner sides of the limbs. The epidermis has two layers or zones; the outer is called the horny zone and the inner the germinative zone.

The *horny zone* has three layers:

1. The *horny layer* (stratum corneum) is the most superficial layer. The cells are flat and have no nuclei and the protoplasm has been changed into a horny substance called keratin, which is water-proof.
2. The *clear layer* (stratum lucidum) is composed of cells with clear protoplasm and some have flattened nuclei.
3. The *granular layer* (stratum granulosum) is the deepest layer. It consists of several layers of cells with granular protoplasm and distinct nuclei.

The *germinative zone*, which is deeper, consists of two layers:

1. The *prickle cell layer* contains cells of varying shapes, each having short processes, rather like thorns, joining them together. The nuclei are distinct.

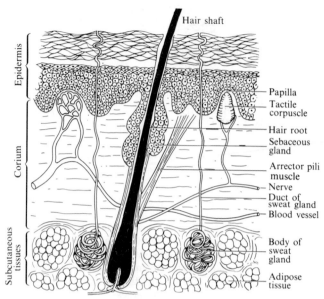

Epidermis

Corium

Subcutaneous tissues

Hair shaft

Papilla
Tactile corpuscle

Hair root
Sebaceous gland

Arrector pili muscle
Nerve
Duct of sweat gland
Blood vessel

Body of sweat gland

Adipose tissue

Papilla of the hair

FIG. 198. Diagrammatic section of the skin.

2. The *basal cell layer* consists of columnar cells arranged on a basement membrane.

The surface scales are constantly being rubbed off by friction and are constantly renewed from below as the deep cells multiply and are driven up to the surface, developing into scales as they approach it. The epidermis has no blood supply and practically no nerve supply. It is nourished by lymph from the blood vessels in the underlying corium. It is the epidermis which is raised in a blister, and the blister can therefore be snipped with scissors without causing any pain, to allow the lymph it contains to escape.

The basal layer of cells in contact with the corium contains the pigments which give the skin its colour: yellow, red or black. The pigment protects the body from the harmful effects of the sun's rays, since dark colours absorb radiation. The scales of the surface prevent the entrance of bacteria into the tissues, since they cannot

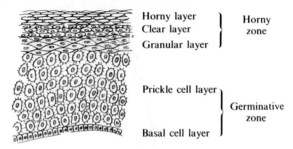

Horny layer
Clear layer ⎫
Granular layer ⎬ Horny zone

Prickle cell layer ⎫
⎬ Germinative zone
Basal cell layer ⎭

FIG. 199. The epidermis.

digest these dried-up cells and make their way through them. Once the epidermis is broken by a cut or prick, infection may enter the tissues and sepsis may follow.

The *corium* is a tough elastic layer which is very thick in the palms of the hands and the soles of the feet and very thin in the eyelids. It consists of connective tissue with elastic fibres, blood vessels, lymphatic vessels and nerves. Numerous conical projections, called *papillae*, extend from the surface of the corium and protrude into the epidermis. They are most numerous where the skin is most sensitive and in these areas they are arranged in parallel ridges, different in each individual, which are useful as finger prints.

The *nerve endings* in the skin are for the most part sensory, and are of different varieties to give the various different sensations of which the skin is capable, namely, the sensations of touch, heat, cold and pain. The nerves of touch end in round bodies known as the touch or *tactile corpuscles* which are stimulated by pressure, and the nerves of heat, cold and pain in delicate, tree-like branches. A few of the branches of these nerve endings pass into the epidermis. Heat is experienced only if a hot thing touches the skin over the termination of a special nerve ending affected by heat. In some parts the nerve endings are so close together that this is not appreciated, but where the nerve endings are less numerous, as over the back of the hand, it is possible to find spots where heat can be felt (hot spots) and others where cold is felt (cold spots).

The arteries supplying the skin form a network in the subcutaneous tissue and branches supply the sweat glands and the hair follicles. Fine capillaries also pass into the papillae.

Appendages of the Skin

The skin carries the following appendages:

1. Sweat glands
2. Hairs
3. Nails
4. Sebaceous glands

The *sweat glands* are twisted, tubular glands which lie deep in the true skin. Their ducts open in the pores of the epidermis, and the tube is coiled up into a little round ball deep in the skin, called the body of the gland. The sweat glands secrete sweat, which consists of water, salts, and a trace of other waste products. Much of this sweat evaporates immediately it reaches the skin surface, and is called *insensible sweat*. When sweating is heavy some of the sweat is poured out on to the skin, which becomes wet with sweat; this is called *sensible sweat*, and evaporation takes place from the whole surface. If sweating is profuse and runs off the body the cooling effect is lost. The secretion of sweat is a means of excretion of waste products, and some poisons and drugs may be got rid of in this manner; its chief importance, however, lies in the fact that the evaporation of sweat uses up body heat, since heat is required to turn water into water vapour. The amount of sweat secreted therefore depends on the amount of heat that the body needs to lose. The average amount excreted in 24 hours is 500 to 600 ml. In hot weather and in violent exercise sweating is heavy, so that much heat is lost by the evaporation of sweat. At these times less urine is passed, so that there is not excessive loss of fluid. In cold weather or rest, less sweat is secreted, so that there is less loss of heat by evaporation; at the same time more urine is secreted by the kidney, so that loss of water is balanced.

Sweat glands are present all over the body, but are larger and more numerous in certain parts, such as the palm of the hand, the sole of the foot, the axillae, groins and forehead.

The *hairs* consist of modified epithelium. They grow from tiny pits in the skin, known as hair follicles. At the base of each follicle is a group of epithelial cells which form the root from which the hair grows. The hair root is the part of the hair within the follicle, extending through both corium and epidermis. The hair shaft projects beyond the epidermis. The hair bulb is the expanded part of the hair within the follicle. At its base is a small conical projection called a papilla which contains blood vessels and nerves to supply the hair.

Hairs are always set obliquely in the skin. The arrectores pilorum are small involuntary muscles connected to the hair follicles. They are always on the side toward which the hair leans so that when they contract the hair stands erect. The skin around the hair is elevated at the same time, which produces the effect known as 'goose-flesh'. The hairs are constantly being shed and renewed. As long as the root is healthy, fresh hair will grow from it, but if the root is destroyed or its blood supply interfered with, the growth of hair will stop. On the scalp baldness results. Good brushing, which stimulates the blood supply, and massaging of the scalp promotes the health and growth of the hair of the head. Hair is present all over the body except on the palms of the hands and the soles of the feet, but is so fine and sparse as a rule that it cannot be seen—a point that should be remembered in preparing the skin for operation. It is longer and more plentiful in the eyebrows, axillae, and groins, where it entangles the sweat, preventing it from running down and assisting its evaporation.

The *nails* are horny plates of modified epithelium which protect the tips of the digits. They grow from a root of typical soft epithelial cells at the base of the nail. This root is embedded in the fold of the epidermis. The nails replace the epidermis here, but are continuous with it, so that there is a continuous barrier to exclude bacteria. It is, however, easy for a break in the continuity to occur here if the nails are not carefully looked after and this is particularly dangerous among nurses, since they are in contact with infection in the course of their work. In animals the nails are protective in more senses than one, but as man has invented implements of various kinds for his use, his nails do not get worn down as do those of the animal, and need regular cutting.

The *sebaceous glands* are small saccular glands which secrete an oily substance called *sebum*. They are situated in the angle between the hair follicle and the arrector pili muscle, so that contraction of the muscle has the effect of squeezing sebum from the gland. This lubricates the skin and hair, and keeps them soft and pliant, so that they do not readily break. However, sebum picks up dust and bacteria, which cling to an oily surface. As a result we must constantly remove it by washing with soap and water. If some substitute is not applied the skin quickly becomes brittle, so that it breaks readily and lets bacteria enter. The nurse, whose hands are often in lotions and soapy water, which remove the sebum, needs to pay particular attention to this point.

Functions of the Skin

1. It regulates body temperature.
2. It secretes waste products.
3. It is the organ of touch and other senses through which we are made aware of our environment.
4. It keeps out bacteria by its dry, scaly outer surface.
5. It secretes sebum.
6. It protects the body by its pigment from the harmful effect of the sun's rays.
7. It produces vitamin D through the action of ultraviolet rays on the ergosterol it contains.

Control of Body Temperature

Body temperature is a balance between heat gained and heat lost. Man is a warm-blooded animal and the temperature must be maintained around 37°C; a rise or fall of one degree or more affects the normal functioning of the nervous system and of the enzymes.

The main temperature regulating mechanism is in the hypothalamus. It acts on a 'negative feedback' system; if the body temperature rises mechanisms come into action so that heat is lost from the body; if the body temperature falls heat is conserved until the temperature approaches normal.

Heat production occurs primarily by metabolic activity. Additional heat is produced by exercise, activity, increased muscle tension, shivering and also by endocrine disorders, infections, trauma and by emotion. Heat production is lowest during sleep and highest during muscular activity.

Heat loss occurs through:

1. Radiation, conduction and convection of heat from the skin
2. Evaporation of sweat
3. Respiration
4. Excretion of urine and faeces

Radiation is the transfer of heat from one object to another without physical contact between the two. The body radiates heat to every object near it and the heat loss is proportional to the surface area; a large area loses a lot of heat but heat loss can be diminished by reducing the surface area. For example, less heat is lost from a body in a curled-up position than in a stretched-out position.

Conduction is the transfer of heat from one molecule to another.

If one end of a metal rod is placed in the fire, heat will be conducted along it until the whole rod is hot. Direct contact will cause the body to lose heat to any object which is cooler than the body.

Convection is the transfer of heat from the body to the air, which then rises and is replaced by cooler air which in turn is heated. Heat loss by convection is reduced by wearing suitable clothing.

Evaporation of sweat from the surface of the skin is continuous and has a cooling effect on the body. It is more effective in dry climates, because when the air is humid and already saturated with water vapour further evaporation cannot occur.

Heat is also lost from the body with every exhalation of air, since expired air contains water vapour which evaporates. Only small amounts of body heat are lost through the passage of urine and faeces.

In *hot* weather to keep the temperature normal:

1. *Heat production* is lessened; the thyroid and suprarenal glands do not stimulate so much tissue activity.

2. *Heat loss* is increased by dilation of blood vessels in the skin so that radiation, conduction and convection are increased and there is increased sweating, so more heat is lost by evaporation.

26 The Reproductive System

The Male Genital Organs

The male genital organs consist of:

1. The testes and epididymides
2. The deferent ducts
3. The seminal vesicles
4. The ejaculatory ducts and the penis
5. The prostate
6. The bulbo-urethral glands

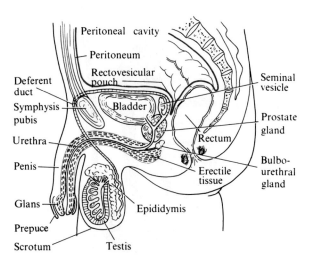

FIG. 200. The male reproductive organs.

The *testes* are the reproductive glands in the male. They are suspended in the scrotum by the spermatic cords but they develop high up in the abdomen close to the kidneys and gradually descend through the inguinal canal into the scrotum shortly before birth. Occasionally one, or both, glands fail to descend and remain in the abdomen or in the inguinal canal; surgery may then be required to relocate them.

As the testis descends it brings down with it a pouch of peritoneum called the *tunica vaginalis*. This forms a serous covering for the testis, but the pouch should otherwise be obliterated. If it is not obliterated it forms a possible site for hernia. This causes an inguinal hernia, a loop of bowel or some other organ falling into the pouch which forms the hernia sac. Each testis consists of 200 to 300 lobules, each containing up to three tiny convoluted tubules called the *convoluted seminiferous tubules*. The epithelial lining of their walls contains cells which develop into spermatozoa by a process of cell division. The tubules are supported by loose connective tissue which contains

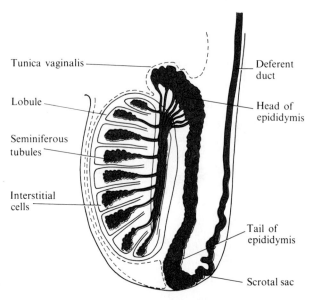

FIG. 201. Structure of the testis.

groups of *interstitial cells*; these cells secrete the male hormone *testosterone*.

The *epididymis* is a fine, tightly coiled tube which is packed into the form of a long narrow body attached to the back of the testis. The seminiferous tubules of the testis open into it and it leads into the deferent duct.

The *deferent duct* is a continuation of the duct of the epididymis. It passes through the inguinal canal and runs between the base of the bladder and the rectum to the base of the prostate gland where it is joined by the duct of the seminal vesicle.

The *seminal vesicles* are two pouches lying between the base of the bladder and the rectum. They secrete an alkaline fluid containing nourishment which forms a large part of the seminal fluid.

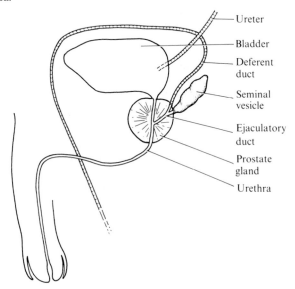

Ureter

Bladder

Deferent duct

Seminal vesicle

Ejaculatory duct

Prostate gland

Urethra

Fig. 202. Relations of seminal vesicles.

The *ejaculatory ducts* are formed by the union of the ducts of the seminal vesicles and the deferent ducts. They commence at the base of the prostate and end at the opening of the prostatic utricle in the urethra.

The *penis* is a tubular organ plentifully supplied with large venous sinuses which can fill with blood causing erection of the organ. It contains the urethra which is common to both the urinary and the reproductive systems in the male. At the tip of the penis is an enlargement called the *glans penis*, in the centre of which lies the urinary meatus. The glans is normally covered by a loose double fold of skin called the prepuce or foreskin. It should be possible to draw the foreskin back over the glans penis, but sometimes the opening in it is too small. This is known as phimosis, and is treated either by stretching the foreskin or by circumcision, i.e. a cutting away of the foreskin, in more severe cases.

The *prostate* surrounds the commencement of the urethra in the male. It is about the size of a chestnut and contains the urethra and the ejaculatory ducts. It consists partly of glandular tissue and partly of involuntary muscle and produces a secretion which is alkaline in reaction (semen) and provides nourishment for the sperm.

The *bulbo-urethral glands* are situated on either side of the membranous portion of the urethra. The ducts open into the spongy portion of the urethra and the glands secrete a substance which forms part of the seminal fluid.

The *seminal fluid* is composed of substances secreted by the testes, the seminal vesicles and the prostate; it contains the spermatozoa.

The *spermatozoa* are minute cells each with a tail-like projection joined to the cell by a constricted portion called the neck. The tail has a lashing movement which enables the cell to move after the semen leaves the male reproductive tract. As a result when the spermatozoa are deposited in the female vagina they can make their

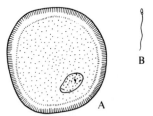

Fig. 203. An ovum and a spermatozöon. (A) Mature ovum; (B) spermatozöon drawn to the same scale.

way up the uterus and uterine tubes in search of the ova. They are produced in enormous numbers and it is estimated that on average 300 000 000 are deposited in the vagina at one time, though only one is necessary to fertilize the ovum.

The Female Genital Organs

The female genital organs consist of an internal and an external group.

Internal Organs

The internal organs, situated within the lesser pelvis, are:

1. The ovaries
2. The uterine tubes
3. The uterus
4. The vagina

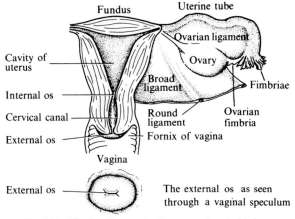

Fundus   Uterine tube

Ovarian ligament

Cavity of uterus

Ovary

Broad ligament

Internal os

Fimbriae

Cervical canal

Round ligament

Ovarian fimbria

External os

Fornix of vagina

Vagina

External os — The external os as seen through a vaginal speculum

Fig. 204. The female reproductive organs from behind.

The *ovaries* are two small glands about the size and shape of almonds, which are situated in the lesser pelvis, one on either side of the uterus, behind and below the uterine tubes. Each is attached to the broad ligament by a fold called the *meso-varium*. The fimbriated ends of the uterine tube and a suspensory ligament are also attached to the ovary.

After puberty the ovary has a thick cortex surrounding a very vascular medulla. At birth the cortex contains numerous *primary ovarian follicles*. After puberty some develop each month to form *vesicular ovarian follicles* (Graafian follicles) one of which usually matures and ruptures, releasing an ovum. This process is called *ovulation*. The ovum passes into the uterine tube along its fimbriated end and may be fertilized by a male sperm. If fertilization occurs it usually takes place in the lateral third of the uterine tube.

After ovulation the vesicular ovarian follicle is converted to a mass of specialized tissue called the *corpus luteum*. If fertilization occurs the corpus luteum remains active until late in the pregnancy; if fertilization does not occur the corpus luteum begins to degenerate after about 14 days. The corpus luteum produces the hormones progesterone and oestrogen; these hormones cause the lining of the uterus, the endometrium, to become thickened, ready to receive the fertilized ovum. However, if fertilization does not occur, the hormones are withdrawn as the corpus luteum degenerates and the endometrium is shed in the process called menstruation (see page 313).

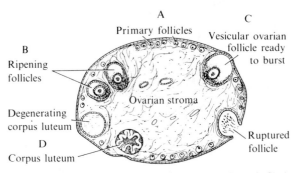

FIG. 205. The ovary, showing: (A) primary follicles; (B) and (C) ripening follicles; and (D) corpus luteum.

The functions of the ovary are controlled by hormones from the hypophysis: follicle-stimulating hormone and luteinizing hormone.

The ovaries begin to function at puberty and continue to discharge ova at monthly intervals from about the age of 13 to about the age of 45 years, the usual time of the menopause. The ovaries are smooth in childhood, but the scars which follow the rupture of the follicles pucker up the surface, so that they finally become very irregular and rather like almonds in appearance as well as shape.

Hormones from the ovaries are also responsible for the development of the reproductive system and the general development which marks puberty in the female, and occurs at about 13 years of age. There is marked development of the external genitals, of the uterus and the breasts: hair grows on the genitals and in the axillae; there is a general rounding of the figure and a gradual development of the typical mental outlook of the female personality, which gradually matures as the reproductive organs themselves mature.

The *uterine tubes* are situated in the upper part of the broad ligaments of the uterus. They are about 10 cm long and transmit the ova from the ovaries to the cavity of the uterus. Each tube has four parts:

1. The *infundibulum* is a trumpet-shaped expansion which opens into the abdominal cavity close to the ovary and has a number of processes called *fimbriae*.

2. The *ampulla* is a thin-walled, tortuous part which forms rather more than half the tube.

3. The *isthmus* is round and forms about one-third of the tube.

4. The *uterine part* passes through the wall of the uterus and is about 1 cm in length.

The uterine tubes have three coats: an outer serous covering of peritoneum, a muscular coat and a lining of ciliated epithelium. The ova are conveyed along the tube by the peristaltic action of the muscular coat and by the action of the cilia.

The *uterus* is a hollow, thick-walled, muscular organ situated in the lesser pelvis between the rectum and the bladder. It is about 7·5 cm long, 5 cm across and 2·5 cm thick and it weighs about 30 g. It communicates with the uterine tubes, which open into the upper part of the uterus, and the vagina, which leads from the lower part. The uterus forms almost a right angle with the vagina, into which the cervix of the uterus protudes. The upper part of the uterus is broad and is called the *body* of the uterus; the part of the body above the entrance of the uterine tubes is called the *fundus*. The *cervix* is narrower and more cylindrical than the body and projects through the anterior vaginal wall. The narrow opening at the upper end is called the *internal os* while the lower opening is called the *external os*. The uterus has three coats; the outer serous coat is derived from the peritoneum, the muscular coat consists of involuntary muscle, the fibres of which are arranged in longitudinal, oblique, transverse and circular layers and the lining is of mucous membrane and columnar epithelium known as the *endometrium*. The *broad ligaments* pass from the uterus to the lateral walls of the pelvis. The *round ligaments* are situated between the folds of the broad ligament below the uterine tubes. The uterus normally lies in an anteverted position with the fundus facing the abdominal wall and the cervix towards the sacrum. There is also some anteflexion, the body being bent forwards on the cervix. The uterus is maintained in this position by the *transverse ligaments* which run from the cervix to the lateral walls of the pelvis and by the *utero-sacral* ligaments which run from the cervix to the sacrum. The uterus is also supported indirectly by the pelvic floor (see Fig. 81).

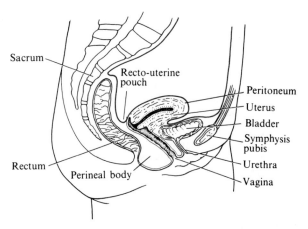

FIG. 206. Median section of the female pelvis, showing the position of the uterus.

If the ovum is fertilized in the uterine tube it embeds in the thickened, vascular endometrium which has been prepared to receive it by the action of the hormones. It remains there, increasing in size, until it fills the uterus after which the uterus grows with it until the end of the period of pregnancy. At the site of implantation the placenta develops; this is the organ through which the fetus

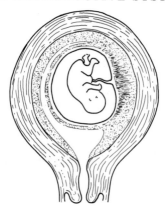

FIG. 207. The uterus at about the tenth week of pregnancy showing the placenta and developing fetus.

receives nourishment and oxygen from the maternal blood during intra-uterine life.

Menstruation is the shedding of the thickened endometrium, with some blood, which occurs each month after puberty until the menopause.

The anterior lobe of the hypophysis secretes follicle-stimulating hormone (FSH) which initiates the development of a follicle in the ovary. As the ovum matures the follicle secretes oestrogen which is necessary for the growth of the endometrium and its preparation to receive the fertilized ovum. Oestrogen is also responsible for the gradual development of secondary sex characteristics in the young girl. When the amount of oestrogen in the blood reaches a high level further secretion of FSH is prevented but the anterior lobe of the hypophysis begins to release luteinizing hormone. Following ovulation (the release of the ovum from the follicle) luteinizing hormone converts the ruptured follicle into the corpus luteum which secretes the hormone progesterone. This hormone completes the development of the endometrium. If the ovum is not fertilized the corpus luteum begins to degenerate, the level of progesterone decreases and the endometrium is shed. The low progesterone level also stimulates the hypophysis to secrete more FSH and the cycle begins again. For convenience of description the first day of the

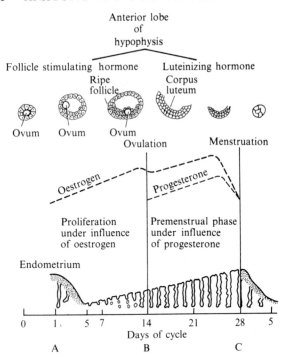

Fig. 208. Diagrammatic representation of the changes in the endometrium in the menstrual cycle.

menstrual flow is designated Day 1 of the menstrual cycle.
Ovulation usually occurs about the 14th day.

The *vagina* extends from the uterus to the labia; it lies behind the
bladder and urethra and in front of the rectum and anal canal. The
cervix enters the anterior wall of the vagina at right angles so the
posterior wall of the vagina is longer than the anterior wall. The
recesses formed by the projection of the cervix into the vagina are
called *fornices*. The walls of the vagina are normally in contact with
each other. The vagina has two coats; a muscular coat which has
longitudinal and circular fibres and an inner lining of mucous
membrane which is in folds or rugae. After puberty this lining

becomes thick and is rich in glycogen; the action of certain bacteria (Doderlein's bacilli) on the glycogen renders the vaginal secretion acid in reaction. The peritoneum passes over the body of the uterus on to the back of the cervix and close to the posterior fornix of the vagina before being reflected back in front of the rectum. This fold is called the recto-uterine pouch.

External Organs

The external genital organs in the female are collectively called the *vulva*. They consist of:

1. The mons pubis
2. The labia majora and minora
3. The clitoris
4. The vestibule of the vagina
5. The greater vestibular glands

The *mons pubis* is a pad of fat covered with skin lying over the symphysis pubis. It bears hairs after puberty.

The *labia majora* are two folds of fatty tissue covered with skin extending backwards from the mons on either side of the vulva and disappearing into the perineum behind. They develop at puberty and are covered with hair on the outer surface after that stage. They atrophy after the menopause.

The *labia minora* are two smaller fleshy folds within the labia majora. They meet in front to form a hood-like structure called the *prepuce*, which surrounds and protects the clitoris. The labia minora unite behind in the frenulum of the labia minora (fourchette). This is merely a fold of skin which is often torn in the first labour. The labia minora are covered with modified skin rich in sweat and sebaceous glands to lubricate their surfaces.

The *clitoris* is a small sensitive organ containing erectile tissue corresponding to the male penis. It lies in front of the vulva immediately below the mons pubis and is protected by the prepuce.

The *vestibule* is the cleft between the labia minora. The orifice of the vagina and the orifice of the urethra open into it.

The *orifice of the urethra* lies at the back of the vestibule, projecting slightly from the normal surface level. At the entrance there are two fine tubular glands, called urethral glands, which secrete lubricating fluid and are more important because they tend to harbour infection in cases of gonorrhoea.

The *orifice of the vagina* occupies the space between the labia

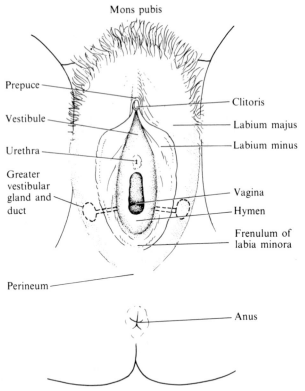

Mons pubis

Prepuce

Vestibule

Urethra

Greater
vestibular
gland and
duct

Perineum

Clitoris

Labium majus

Labium minus

Vagina

Hymen

Frenulum of
labia minora

Anus

FIG. 209. The female external genital organs.

minora behind the vestibule. It is normally a slit from front to back,
the side walls of the vagina being in contact. The orifice is largely
blocked in the virgin by the *hymen*. This is a double fold of mucous
membrane, usually crescent-shaped, and leaving a gap at the front
for the escape of the menstrual flow. Sometimes it has a number
of small perforations; this is called a fenestrated hymen. Sometimes
there is no opening, but this is usually owing to the vagina not
having developed into a canal. The hymen is placed a little inside
the orifice of the vagina, so that there is a slight depression between
the frenulum of the labia and the hymen.

The *greater vestibular glands* are two small glands lying one under each labium majus on either side of the vaginal orifice. Their ducts open laterally to the hymen. They secrete a lubricating fluid to moisten the surface of the vulva and so facilitate sexual intercourse.

The whole surface of the vulva is covered with modified skin, i.e. stratified epithelium. It carries hair on the outer surfaces only, but the inner surfaces are exceptionally rich in sebaceous and sweat glands, so that the surfaces are moist and there is no friction in walking.

The *perineum* is the expanse of skin from the vaginal orifice back to the anus. It is about 5 cm in length and bears hair. It lies over the *perineal body*, a mass of muscle and fibrous tissue separating the vagina from the rectum. The muscle of the perineal body is largely the levator ani, the chief muscle of the pelvic floor.

Breasts

The breasts are two glands which secrete milk and are accessory organs to the female reproductive system. They are present in rudimentary form in the male. They lie on the front aspect of the thorax

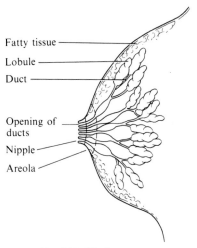

Fatty tissue

Lobule

Duct

Opening of ducts

Nipple

Areola

FIG. 210. The breast.

and vary considerably in size. They are circular in outline and convex anteriorly. In the centre of the surface is the nipple, which projects normally from the skin level and is pink in the virgin, but pigmented after the first pregnancy.

The breast is a compound saccular gland with its ducts converging to the nipple and opening on its surface in large numbers. The gland tissue is similar to the tissue of the sebaceous gland, but more highly developed and secreting milk in place of sebum. The gland is divided into lobes by partitions of fibrous tissue—a fact which makes the draining of an abscess in the breast difficult. The lobes are subdivided into lobules.

The breasts develop at puberty under the influence of hormones, and further development occurs during pregnancy as a result of hormones from the pituitary gland and ovaries. Fluid known as colostrum is secreted in small amounts by the gland during pregnancy and at the time of childbirth, but the true milk is not secreted till about the third day of the puerperium or lying-in period.

Further Reading

Avers, Charlotte J. (1974) *Biology of Sex*. Chichester: John Wiley & Sons.

Beck, Mary E. (1977) *Nutrition and Dietetics for Nurses*. Edinburgh: Churchill Livingstone.

Block, Bartley C. (1975) *Man, Microbes and Matter*. Maidenhead: McGraw-Hill Book Company (U.K.).

Calder, Nigel (1970) *The Mind of Man*. London: BBC Publications.

Clarke, C. A. (1970) *Human Genetics and Medicine*. London: Edward Arnold.

Freeman, W. H. and Bracegirdle, B. (1966) *Atlas of Histology*. London: William Heinemann Medical Books.

Gibson, Jenifer M. (1977) *Modern Microbiology and Pathology for Nurses*. Oxford: Blackwell Scientific Publications.

Gibson, John (1974) *Guide to the Nervous System*. London: Faber and Faber.

Gregory, Richard L. (1972) *Eye and Brain*. London: Weidenfeld (Publishers).

Hare, R. and Cooke, E. M. (1975) *Bacteriology and Immunity for Nurses*. Edinburgh: Churchill Livingstone.

Kilgour, O. F. G. (1969) *Introduction to the Physical Aspects of Nursing Science*. London: William Heinemann Medical Books.

Kilgour, O. F. G. (1972) *Introduction to the Chemical Aspects of Nursing Science*. London: William Heinemann Medical Books.

Mason, A. S. (1960) *Hormones and the Body*. Harmondsworth: Penguin Books.

Munro, J. M. (1972) *Pre-Nursing Course in Science*. Edinburgh: Churchill Livingstone.

Rose, S. (1970) *Chemistry of Life*. Harmondsworth: Pelican.

Ryan, B. and Keen, J. (1970) *Basic Science for Nurses*. Maidenhead: McGraw-Hill Book Company (U.K.).

Tighe, J. R. (1972) *Pathology*. London: Baillière Tindall.

White, R. G. and Timbury, M. C. (1973) *Essentials of Immunology and Microbiology*. Tunbridge Wells: Pitman Medical Publishing Company.

Williams, P. L. and Wendell-Smith, C. P. (1969) *Basic Human Embryology*. Tunbridge Wells: Pitman Medical Publishing Company.

Index